RUMINATIONS FROM A SEASONED CURMUDGEON

Essays gleaned from the fruits of observations of human nature throughout a lifetime of many & varied experiences.

by Neely E. Pardee, M.D.

Edited by Ellen Pardee

So much to ponder, so little time...

EP

DISCLAIMER

This book is designed to express the conclusions and opinions of the author. The contents are solely the products(s) of the author's study and work and have not been aided or abetted by any third party. These thoughts and conclusions should in no way construed as advice or as a substitute for the reader's own thought processes. As the author is responsible for his own reactions to his writing, so is the reader for his.

COPYRIGHT NOTICE

TABLE OF CONTENTS

FORWARD

I met Dr. Neely Eugene Pardee when I was twelve. He was 'Ellen's Dad' to me at first (Ellen being my best friend), then 'Mr. Pardee' until I learned to address him as 'Dr. Pardee'. Now, almost five decades later I have advanced to calling him 'Gene' and sometimes slide into 'Dad' because he really is like a Dad to me. You could say I've spent a fair amount of time with him.

Oh, the times & the things I've experienced in his presence - my first sighting of phosphorescence, my introduction to the outhouse, the first time I played charades, first time (followed by hundreds) to partake in the tradition of hors d'oeuvres, hiking & snorkeling adventures, exposure to music of all genres from Alice Cooper to Beethoven, Glen Miller to Gershwin, and sonatas to sea chanties, the wise toast at breakfast of 'good health, good luck & be happy', eating out at a real restaurant not called McDonalds, philosophical and logical discussions and explorations of topics like 'do flies sleep' - 'what's the difference between a turtle and a tortoise' & 'why do some people sneeze once-others up to 7 times and how do they even do that?'.

Gene will have an assorted, motley, multifarious collection of books at any and all times in the 'book bag' that goes everywhere with him that's anywhere longer than a day from home, because, of course, at home there are walls of books. In the bag you can find anything from Ayn Rand, Darwin, The Bible, Umberto Eco, Alexander McCall Smith, C.S. Forester, and so many more. Never, never, never is he far from a Dictionary, Random House Webster's College Dictionary to be specific (no thank you, Internet & Google). Oh, there was one time 'the dictionary' was unaccountably left behind on a vacation to St. John) and it was often and very audibly lamented.

Curiosity. I think all humans are born curious. I think we have to be in order to survive because survival depends on learning by exploring our curiosities - thus we learn to survive. Once surviving, our curiosities can really be put to work to eclipse levels that answer all arising questions and interests that life's time allows. Gene is human, was born 'curioser' than most and I swear he has more memory cells ("knowledge, after all, is memory," as he says) than the 'above, above average' person - combine all of the afore-mentioned and you get the genius of Gene. At times when I've watched him think my mind puts special effects in (like you would get in a new Sherlock Holmes movie) with the zip, crack - thought bubble, pur-vip, snap - thought bubble, whiz wham - conclusion. I can just see the 'think' going all over the place in his brain as he closes his eyes, tilts his head up and rests his hands perfectly still in his lap. Yet, he enjoys melting an Oreo in his mouth - yes, really & I think he has timed and studied this phenomenon too.

His favored dictionary defines curiosity as 'a state in which you want to learn more about something'. He has the disposition to inquire, investigate, seek knowledge & observe to satisfy his desire to gratify his mind with new information & articles of interest.   Of course in doing this he forms an opinion culminating in a conclusion - all of which will be shared with you in this collection of 'ruminations'. So, as he states in one of those ruminations, 'It is evident that this set of ruminations, which began with an interest in exploration, is itself an exploration'. One of my favorites is 'Us and Them' because it explores belonging, how it matters & contributes to the sense of self. Well, I'm glad I belong to the group called 'Pardee' & boy did it contribute to the kind of person I grew up to be - not really so bad at all. Read on and Gene will take you through the exploration of his curiosities and you could realize some of them are yours as well!

Sue Robertson
October 2014

PREFACE

These essays began as an exercise to determine whether anything cohesive or universal could be mined from the infinitely available observations of people, in and out of their personal worlds. An occupation that requires the most accurate of observations in order to be able to make crucial decisions and a temperament just as demanding together have gathered an enormous harvest over a lifetime of these observations. Once retired the author's observations could not and did not stop but they certainly needed a place to go. Thus, with plenty of time and a desire for mental effort, the careful study, a glimpse of an interesting tableau, and many snippets overheard or briefly mentioned transformed themselves into the study of we humans as a group, in groups and individually. In a unique way these essays bring the scientific side and psycho/social side of human behavior together to achieve a better understanding of our many-faceted humanity.

These essays are not meant to persuade you of a certain point of view; neither do they intend to tell you, and expect you to believe, that the definitive answers to physical and social behaviors lie within them. What is hoped is that you will choose to think, to ask questions of yourself and others, to allow your curiosity full rein.

For not only is it fun (see Webster's Random House Dictionary; *4. Providing pleasure or amusement; enjoyable*) as well as gratifying to believe you may have or have come upon some answers, it can be fun and gratifying to be searching for answers to questions old and new. There are a variety of subjects treated in these essays as a result of having an open mind that allows itself room to roam. In the end, though, all the scattered thoughts, the newly learned facts and figures, and the observations create that special urge to set it all down and uncover the results. With those results, another quest can easily be found. Maybe you can find a quest or two here.

EP, November 2014
Kirkland, WA

COMPETITION

THE WORD

The first phrase used to define the meaning of the word "competition" in the Random House Webster's College Dictionary is, "The act of competing; rivalry for supremacy, a prize, etc.: 'competition between two teams.'" The other four expressions which follow the first expand on this notion (or these notions), that competition involves a contest between two or more individuals for the obtaining of some reward, or the reaching of some goal. What we refer to as competitions have been parts of human life forever. All of the implications of the use of this word are not included in its dictionary definition, nor do all of them seem to be thought of consistently by people engaged in competitive activities. The aim of this essay is to explore some of those further implications.

TYPES OF COMPETITORS AND COMPETITIONS

In any contest there are two or more entities simultaneously attempting to reach a goal which only one, or a few out of many, can achieve. Some competitions are between members of the same biological species and some pit members of different species against each other. A young stag may fight with an older one for dominance over a group of does, a young person may compete with his or her parent in order to prove that he or she is more accomplished or in some way superior to the elder. Inter-generation competition is, or may be, an example of a competition in which the competitors are not necessarily competing directly. A competitor may be directing his actions to defeat or in some way dominate an adversary who exists, during the struggle, only in his mind or imagination. A boy with little athletic talent may be unhappy when he is not chosen to be part of a football team and thus has been denied the pleasures of the actual playing and of group acceptance. He may then think to himself, "they don't want me on the team, but I'm better than any of those guys at math." Those to whom he is comparing himself do not see him as an adversary and do not realize he is thinking of them in that light.

Groups compete for prizes, titles, food supplies, hunting

grounds, sheltering habitats, dominance and other rewards. These groups may be whole species, herds of animals, residents in or of neighborhoods, religious denominations seeking to recruit converts, political parties and advocacy groups, or organizations like those campaigning to have particular subjects included in school curricula or business organizations. Evolution is in many ways a set of competitions, both within and between species, although non-competition factors like climate change are also in play.

The building of houses and the development of subdivisions on previously undeveloped land has some of the quality of a competitive activity, as wild animals' habitats are taken from them. Individual molecules in the fluid around a cell surface, each one not an individual organism, compete for attachment to binding to receptors on the cell's surface. There are more molecules than receptors, so not all of them can find a binding site. This situation illustrates a fundamental aspect of competition: in any contest, there are more participants than winners.

The notion of goals is inherent in that of competition. There is no contest, if two or more individuals are not attempting to reach the same goal. The goal or goals of a competition is (are) such that all the involved competitors may not reach it. At the end of any competition there are usually one or more winners and one or more losers (except when the competition ends inconclusively). Further, all the competing individuals agree about what the goal is and that it is worth reaching. Contestants also ordinarily agree on the rules competitors must follow, whether they are explicitly formulated, implicit or dictated by the context and circumstances of the contest. In baseball the rules are explicit; in hand-to-hand combat, those struggling agree, without having to express their agreement in words, the contest will continue until one of the participants is dead, disabled, taken prisoner or surrenders. This agreement extends, as indicated by this last illustration, to what defines the end of the competition, when one or more competitors reach the goal.

Competitions may begin spontaneously or they may be formally planned and organized. A hiker finds that a sandwich he has brought with him to eat on his hike does not taste as good as he had expected. He throws it away after taking a bite of it. Two dogs nearby see the food fall to the ground and both start to run toward it.

Some competitions are forced rather than being chosen by the participants. These should be distinguished from spontaneously occurring contests. The dogs running to get the discarded food could stop racing if the food fell into a lake or over the edge of a bluff and was then unobtainable (the dogs make a choice); the members of a tribe competing with another for control of a hunting ground must continue their efforts, or they may starve (no real decision to make).

Structured competitions are integral parts of the everyday lives of most of us. They are involved in individual and group survival, well-being and pleasure. (The idea that an individual's experiencing some pleasure periodically may be necessary for his survival comes to mind). Losers, or losing groups, in formal competitions (athletic teams, and so forth) commonly survive, in contrast to what happens in hand-to-hand combat, although they may have feelings of disappointment or frustration. Beneficial effects of competitions include problem solving, as when two manufacturers compete to be the first to eliminate an undesirable feature of a product both of them make and sell, so as to attract more customers. Positive consequences of competitions which are formally organized and carried out are the discharge of individuals' competitive urges in non-destructive ways, so that physical combat is avoided (an election produces fewer casualties than an armed revolution) and the providing to persons not directly competing (sports fans, supporters of political candidates) opportunities to share the rewards of winners vicariously.

We see, and often engage in, informal, sometimes unplanned, competitive social activities. In most instances these seem to be directed at earning praise, provoking envy, enhancing self-approval or increasing a contestant's sense of having power or authority. Examples of this kind of behavior include driving aggressively, to get ahead of other cars on a highway (or any other action directed at "getting there first"); trying to be the best-dressed person at the party; and working to have the most-admired house and yard. Informal competitive behavior in conversations may involve a person's trying to "have the last word," to "say it better" or to be seen as the one who "has it right," in contrast to other individuals whose statements come to be seen as incorrect or incomplete. This last type of competition may be formally organized instead of being coincidental and spontaneous and includes court actions and the actions of the members of debating societies. The occurrence of these organized competitions must depend on and make use of consensus, by competitors or potential competitors, concerning how it is decided when a contest is won, who can or does win it and what the rules are which govern the competitors' behavior. Informal, social and spontaneous competitions do not seem to have explicit "rule books." At times, of course, social competitions may have written rules, as when a fashion designer declares that certain styles are in vogue. At other times, as with competitive driving or arguments arising during a conversation, those competing must have similar beliefs about what constitutes winning and how they should behave. These beliefs may be formally recognized (published logical principles may be referred to during an argument) or unconsciously recognized by the contestants (who had the last word "that time").

Interactions between family members or neighbors often have a competitive quality as with sibling rivalry or attempts to have the neatest, most attractive yard on the block. Sibling rivalry is an example of the principle that the rewards coming to the winner or winners of a competition can be obtained by only one, or only a few of the competitors. There is a limited number of minutes in a parent's day, during which he or she can interact with a child. If one, or a few, of the children in a family gets all of this time, the loser or losers get none. "Mother loves me more than you" (this effect is, or course, a different kind of rivalry than that in which siblings compete to determine which is the stronger or the more skillful in sports activities). Exclusions or exclusivities in friendships and romantic relations have some of the quality, "she's my best friend, not yours!"

WINNERS' REWARDS

Winners' rewards and competitors' goals are closely intertwined. Rewards fall into two broad classes. The first is made up of tangible rewards: trophies, applause, the avoidance of rejection, and acquisition of leadership status, for instance. The reward of leadership status deserves special mention. It carries with it the acquiring of power or authority. A leader can imagine a goal and act to achieve it with less hindrance than would be the case if he were not a leader. This can be in the doing of something he "just wants to do" as an individual, without being told he has no right to do it, and also in having the power to direct others to work toward an objective (Piaget mentions "the well-known joy of being the cause"). Having power in general, and power over others, is clearly a highly valued goal, whether it is power over other members of a family or over the citizens of a nation, considering the lengths people go to get and retain power. Dictators do not give up their roles easily. Only one Roman emperor retired voluntarily (Diocletian); all the others were involuntarily retired by insurrection or assassination. Both a leader and those he leads (who may or may not have been competitors in the contest which resulted in the leader's acquiring his role) may win, in a sense. The general wins a battle and the soldiers he commands win also. The subjects of a wise and clever ruler benefit from the way he rules.

Winning brings applause and approval: the cheers of the fans of a winning team; the invitation to join a social group; the qualifications for membership in which include winning in some competitive activity (Phi Beta Kappa membership, for which students qualify by having high marks in their college courses); and the avoiding of rejection (recall the boy who was not chosen to play on the team).

The second type of reward is invisible, and is individually experienced by the winner or by a person who shares the winner's experiences vicariously. The winner feels good, having won. A wanted, positive feeling state appears in his field of attention, in consciousness. Both logic and personal experience suggest that this second-order reward, or the prospect of its being earned, is the fundamental goal contestants seek, though they may not think consciously about this. A winner may feel emotional rewards like elation, relaxed contentment, joy or freedom from anxiety. Relief of anxiety has features which suggest it may have two components. First the pleasurable recognition that a threat has disappeared; then an increased sense of well-being which appears in consciousness (and persists there for a time). It is as though relief of anxiety has caused stimulation or disinhibition of a feel-good holding and sending domain. There is good evidence that such feel-good sending systems exist. A rat which learns that by pressing a lever in its cage it can activate a stimulating electrode placed in a special location in its sub-cortical fore-brain acts as though this pressing causes it to feel pleasure, neglecting opportunities to eat or copulate and abandoning its usual grooming activities. Considering the ways people behave, it seems likely that such pleasure centers exist in the human brain also. The fact that emotions besides pleasure or well-being do not appear to have direct connections to externally generated sensory input (taste, touch, warmth and so forth) argues that other emotions, besides pleasure or satisfaction, come to consciousness when their specialized holding-sending systems or domains send signals when they have been stimulated. This could account for the fact that we experience mixtures of emotions, pleasure blended with excitement or with relaxation, for example.

It is clear that feeling-state sending domains, each of which does not send its signals with sufficient intensity to consciousness for its particular emotion to be perceived all the time, must be activated by something, with any particular reward. It is also clear that activating signals must arrive at emotion-sending domains from somewhere else in the nervous system. Our daily experiences point to the conclusion that activating stimuli must travel along association chains which have themselves been caused to operate by some trigger, necessarily an external event stimulating sensors (sight, touch, etc.) or cognitive activity ("this is great! I've finally figured it out!").

Functional domains from which emotions are projected toward the attention field may or may not have easily definable anatomic correlates. Some of the reward-sending systems are clearly at least partly localized, as with the rat's sub-cortical "pleasure center." Others, or other parts, of the complex of sites and systems which send feel-good signals to consciousness may have widely scattered components. Cells with opioid receptors at their surfaces probably represent this class. These cells appear to be especially concentrated in some parts of the brain (see Kandel, et al), but are not strictly localized into specific centers. Arrival of morphine molecules at, and the binding of these molecules to, opioid receptors leads to the cell's response, to its sending out a signal which, causes the individual who has received the drug to experience pleasure, well-being or euphoria, in addition to having relief of any pain he may have had.

Memory, or the collection of memory bins, is another functional domain. Memory bins, by themselves, when stimulated, do not appear to be able to send feeling state signals to the attention field. The signal output of a memory bin can, however, bring to consciousness an emotion if its signals, travelling along an association chain, act as stimuli to a feeling-state sending domain. "When I saw her just now, across the room, I remembered how agreeable it was talking to her the other day."

Feeling-state, emotional reward events must occur when input from primary sensors (taste buds, touch sensors in the skin, the retinas and so forth), or from the results of some cognitive activity, causes stimuli to move along an association chain, so that they reach some feeling-state generating domain. Which emotion-projecting site a signal reaches must be determined by a (usually unconscious) screening function, which directs signal movement according to what is in an individual's memory, how particular memories relate to inborn response tendencies (the urge to explore the unfamiliar, or to act to make pleasurable experiences last) and to what associations particular memories have had with what emotions. The experience of the individual who recalls a pleasant conversation is a case in point.

The feeling-state rewards for winning can be experienced by people who themselves have not actively competed. Sports fans, theater-goers, and novel readers feel some of the same emotions as the members of the teams they favor, of the characters they see on stage or on-screen or about whom they read. The person who comes to the end of a story and reads "and they lived happily ever after" is likely to feel (vicariously) some of the emotions of the story's characters.

VARIATIONS IN COMPETITIVE URGES

There is clearly great variation in the intensity and expression of competitive urges, depending both on the setting of a competition and on the natures of the individuals involved in it. We do not often see the members of a herd of cows fight for the right to graze in a particular area of a meadow (in this case there is plenty of grass available, of course). On the other hand we often can see a seagull chasing other gulls away from a piece of food it has started to pick up. Individuals of the same species, in many ways alike but in other ways diverse (size, skin color, physical strength and so forth) vary greatly in the apparent intensity of their competitive urges and how they express those urges. The circumstances in which people find themselves (whether by accident or design) also influence the drive to compete and whether or not they decide to compete in the first place. When food is abundant, there is no incentive to compete to obtain a fair share of it (the way some people strive to accumulate wealth when they cannot possibly spend more than a small part of what they have gained is another question – perhaps related to the development of a sense of having power). The seagulls that might compete can be seen to be eating peacefully side by side on a beach which offers an abundance of their preferred food.

Some of the variations in competitiveness come from individual differences in personality and temperament. Others are due to experiences the different potential competitors have had. "I could not force myself to try that hard," and "I tried before to do that, and I just couldn't" influence how we feel about a competition we may be thinking of entering. We also make judgments, based on experience and on our particular interests and values, about the rewards possibly coming from winning, and the penalties likely to be suffered because of losing a particular contest. "It just doesn't seem to be worth the effort to do that," or "If I could get one of those, I'd be really, really happy!"

The ages of potential competitors have effects on whether episodes of competition begin, and on how they come to their ends. For some people, the urge to compete may decline with age. For most people, the physical abilities and emotional drives needed for competing diminish over time. There are few professional football players over age 40, and there are many people, whose work may have been, in one way or another, competitive, who retire at the ends of their careers and who appear to be content. Sports teams are ordinarily made up of individuals who are near the same age. In many such instances, of course, this is because the teams are school or college-based. It is unusual for teams the members of which are youthful to compete against those composed of elderly individuals.

In any current or anticipated contest a person's knowledge, his own values and abilities, and those of others with whom he is competing or may compete, influences his decision to enter the contest, or to continue in it if it is ongoing. If he feels that he is, or is likely to be, overmatched by many or most of the others in the contest, he may hesitate to begin or may drop out of it. On the other hand, if he says to himself, "I can win that easily. Those guys don't have any idea about how to get it;" his sense of his capacities in comparison with those of others may encourage him to enter, or to continue in, a competitive activity (in this latter case, however, a person's beliefs about his abilities may actually work against him. Overconfidence may cause him to reduce the efforts he makes, and this can diminish the likelihood that he will win).

The number of competitors in a contest also influences competitive drives. A contest with only five competitors has a different quality than does one with 500. When only five contestants are involved, each one is likely to know, or to learn quickly, the other contestants' abilities and personalities. This is not the case when hundreds of individuals are concerned, as in competitions for scholarships. With groups, the effects of competing groups having different numbers of members is ordinarily taken as a given. A large army will defeat a small one if the two are otherwise comparable.

Group membership has some additional effects on members' competitive drives. The sense of group membership can depend on the different ways of defining groups: ethnicity, common acceptance of a religious or political belief system, occupation or geographic area of residence ("I'm a New Yorker"), to name a few. The crusades aimed to displace the non-Christian rulers of Jerusalem; we are reminded daily in news reports of the black-white divide; strikers demanding higher wages are opposed by the calling in of strike-breakers by management. In each case the attitudes of group members towards their opponents might be expressed by, "They're not people like us. They don't deserve to (or have a right to) win." This power of group membership in molding competitive drive is demonstrated over and over. Committed members of groups take great risks, work to the limits of their powers and even die, for what they see as the good of their groups, witness martyrdom in its various forms (including the acts of suicide bombers).

FURTHER DISCUSSION, WITH NOTES ON NEGATIVE EFFECTS
OF COMPETITION, SUMMARY AND CONCLUSIONS

Several characteristics are common to all competitions. In each contest there are two or more persons or animals with the same goals and only one, or only a few out of a large number of individuals, can reach that goal. This goal is seen by all concerned as desirable and potentially achievable. Also, there is implicit or explicit agreement among all competitors about what rules govern their actions and about what defines winning. In some cases the principles which determine what contestants do are inherent in the nature of the competition and its setting. In a battle, what the opposing armies' resources are for obtaining food and ammunition, what the terrain is like over which the battle is fought and on what the weather conditions are during the combat, affect the nature and outcome of the battle. There is no prior agreement between the generals about how the combat is to be conducted, (In some historical situations what combatants have done has been influenced by what were considered to be the rules of war, which included rituals or procedures concerning how the surrender of the losing army was recognized and how the winners should treat those who were defeated). Battles between armies are in substantial contrast with athletic competitions, in which the rules are explicit and have been established in advance of the game or race.

It is clear that competition is a component of many of our activities. It is important in promoting the survival and well-being of individuals and groups. Competitions of one sort or another occur in all cultures. All of us are involved in competitive activity at some points in our lives. The pervasiveness of competition argues that competitiveness has a genetic component. Studies of behavioral tendencies in twin pairs, when identical twins are compared with fraternal pairs and with the broad general population, support this notion.

The rewards which winners earn fall into two broad categories. The first is that of tangible and observable benefits: trophies, titles, applause, food, the gaining or maintaining of group membership, and power, as when a leader establishes his role or an individual gets the freedom and the resources to do the things he wishes. A winner who experiences these rewards commonly also gains one of the second kind – emotional ones: joy, pleasurable excitement, feelings of being wanted or valued, relaxed contentment, freedom from anxiety or a combination of wanted feelings. Winning acts as a trigger which initiates a process which begins with the winning experience and the perception of its outwardly observable rewards and then leads to the activating of an association chain in the winner's brain, along which, or at the end of which, stimulant signals are presented to the pleasure center, or centers, from which characteristic signals are sent to consciousness. The occurrence of such a sequence can be thought of as a reward event. We do not feel especially happy, above the usual baseline sense of well-being, all the time. Winning a contest brings with it an increased feeling of satisfaction or pleasure, which typically lasts for a time beyond the first holding of the trophy, and which can recur with a later reminder of the winner's success (it may last more than momentarily then, also). It is as though the feel-good-sending domain or domains tend to continue transmitting their signals to consciousness, once they have been stimulated, with gradually diminishing intensity.

Competitive events have both positive (wanted) and negative (unwanted) properties and results for competitors, fans, spectators and readers. Among the unwelcome effects of competition are death, physical injury, disappointment, poverty and the disappearance of national, racial or cultural groups. These negative effects are ordinarily suffered by losers, although winning groups or individuals may be so damaged as a result of the intensity of effort they have expended in achieving victory that they are altered in important ways or made weaker by the conflict they have apparently won. A wrestler might succeed in "pinning" his adversary, at the same time suffering a severe sprain which might partially disable him.

Some contests may be prolonged to the point at which the energies and resources of all competitors are exhausted. Such competitions end inconclusively, without there being any real winner. A war could end when both of the opposing armies used up their food and ammunition at about the same time, so that neither could continue to fight. Some long-lasting non-violent competitions leave important issues unsettled, as in actions carried out in legislative bodies, like the United States Congress.

Competitions may be affected in how they are carried out and in who wins if one or more of the competitors fails to follow the assumed or explicitly formulated rules thought to govern it at its beginning. A student cheats on an examination; a college football team secretly adds one or several players with special talents to its roster. This concern applies to formally organized contests, and not to such context-formed conflicts like hand-to-hand combat between soldiers in opposing armies, when the only rule is that there are no formal rules. With formally structured competitions, deviation from rules can make it impossible for them to be carried out in a satisfactory way. If a chess player can move any of his pieces anywhere he wishes, anytime he wishes, it becomes impossible to tell who wins and who loses. Another perplexing question involves what an up-to-now honest player may do if he discovers his adversary has been cheating. Does he break off the game? Does he start to cheat himself? There may be no simple answers to these questions.

The beneficial effects of competition are both numerous and diverse. Among these are the discharge of potentially harmful or destructive urges in organized contests like football or basketball and actions in courts of law. Transfers of leadership roles from one person to another by elections lead to many fewer deaths and physical injuries and to much less disruption of a nation's economic activities than does a violent revolution. Competition leads to the development of group structure when it establishes who the leader is and how leaders and those who are led play their roles. Some non-leaders, who may have competed at some point with the person or persons who achieved leadership status, and thus who are, or have been in a sense, losers, may ultimately benefit from the competition, if their leaders are sensible and competent. The subjects of wise kings have been contented and satisfied with their rulers.

Competitions have resulted in a number of additional benefits. They have led to innovations and to new scientific discoveries (researchers and engineers try to outdo one another), to the solving of social and political problems (government figures seek praise or approval – and re-election – by putting in place needed laws) and to development of new products and services by manufacturers and business organizations competing to gain customers and increase their profits. However, it is clear that the motives driving many human activities are complex, due to forces other than the competitive urge in many cases. In many situations, expectations of other satisfactions occurring as results of action like simple sensory pleasure (good tasting food, orgasm, and so forth) and the previously mentioned "Joy of being the cause" drive what individuals do, and competitive urges simply modify or add force to these tendencies.

SOURCES

Random House Webster's College Dictionary, Random House, New York, 1991

The Dictionary of Psychology, Corsini, R, Editor, Brunner-Routledge, New York, 2002

Principles of Neural Science, Fourth Edition, Kandel, E.R., Schwarta, J.A., Jessell, T.M., Editors, McGraw-Hill Health Professions, New York, 2000

Caplan, F., The First Twelve Months of Life; Your Baby's Growth Month by Month, Bantam Books, New York, Originally published as 12 booklets in 1971, Grosset and Dunlop Hard-cover edition, May 1973

An Encyclopedia of World History, Revised Edition, Langer, W., Editor, Houghton Mifflin Co, Boston, 1904 and 1948

Piaget, J., Play, Dreams and Imitation in Childhood, W.W. Norton and Co, New York, 1962

Bronson, P., Merryman, A., Top Dog: The Science of Winning and Losing, Twelve, Hatchette Book Group, New York, 2013

Bandura, A., Social Foundations of Thought and Action, a Social Cognitive Theory, Prentice Hall, Upper Saddle River, NJ, 1986

Print Media
a. The Seattle Times newspaper, The Seattle Times Co, Seattle, WA

b. The New Yorker magazine, Advance Magazine Publishers, Inc., copyrights through Conde Nast

The common experiences of daily life

PRINCIPLES

SYNOPSIS

Principles are of two general types: "is" principles which are statements about what exists, how things have come to be, and how this world works; and "should" principles about what is a desirable state of affairs or how to reach some goal (here "should" includes "how to" rules concerning action sequences and procedures). Principles, for us, exist in some kind of memory, including the memories of individual persons, words on a page or data stored in computers. Principles have histories. They exist, for us, when they are used in some way. They may disappear or change if they are not used or if the circumstances in which they were first developed change. Principles can be invalid, if the evidence on which they are based, or the reasoning which gave rise to them, are faulty. Bad data leads to false conclusions. Principles may conflict. "Is" principle conflicts usually (not always) disappear with research or increase in general knowledge. "Should" principles' conflicts may last and last, as with Christian-Muslim differences of opinion. Some conflicts can be due to simple misunderstanding. We must have principles and rules if we are to survive and thrive. Fortunately, we seem to have, as a species, an inborn urge to discover and formulate (and record) them.

The Random House Webster's College Dictionary (1991) provides eleven defining phrases for the word principle. The first reads, "An accepted or professed rule of action or conduct." The second states that a principle may be, "A fundamental law, axiom or doctrine." These two definitions point to two different aspects of what we mean when we speak or write about principles: what should be the state of affairs or what should be done in a particular situation, including how to perform actions directed at some goal; and what is the nature of that which exists or how what exists came to be. The other nine defining phrases fall generally into one of these two classes: "should" principles, including "how to" statements; and "is" principles derived from observations by individuals, scientific research and reasoning directed at interpreting the facts discovered in research, exploration, or daily experience.

The first definition usually applies to principles related to people's actions or thoughts ("he really should not have done that."), or to particular conditions. It is inherent in "should" principles that there are, or have been, choices about what persons might decide to do, or alternatives with respect to how something is made or organized. A person can decide not to say his usual daily prayers.

"Should" principles include both statements concerning our behavior and routines or procedures to be followed in pursuit of a goal (assembling a piece of machinery, solving an arithmetic problem). In fact, "should" principles essentially always relate to a goal of some kind. Principles meant to guide behavior are used in preventing interpersonal violence, for example, or in enabling a religious believer to achieve eternal life, besides succeeding in such activities as driving a screw into a piece of wood or frying an egg.

The second defining phrase relates to how things are made, of what they are made and how they came to be. Ice is cold and hard; the universe in its present form began with a "big bang." Such principles are formulated as results of experience and experiment. We discover some in the course of everyday life (like the properties of ice) and some through experiment or study.

"Should" principles necessarily appear, or are developed and recorded as consequences of "is" principles' being perceived or believed. A person trying to drive a screw into a piece of wood "should" rotate it in a certain way, because of how the threads of the screw are ("is") configured. We should pray regularly because there "is" a god and he wants us to do so. Here it should be kept in mind that our understanding may be incomplete or wrong. The amount and nature of the evidence on which the development of an "is" principle has been developed may be limited; the observations on which development of the "is" principle depends may be made in an inappropriate manner, or our reasoning from those observations may be faulty. We are first exposed to experiences which produce "is" perceptions and "is" to "should" inferences as infants and children. Powers outside ourselves (parents, teachers, and other authorities) provide us with smiles, caresses, sweet treats and opportunities to experience pleasure in playing when we do things of which they approve. The development of "is" ideas has much of the quality of classical conditioning, at least in infancy (see Kandel).

It does not matter whether or not God exists and has a quasi-parental relationship with us as long as we believe that he "is" and is concerned with what we do. Our expectations that what we do will lead to particular consequences are potentially based, at least in part, and in some contexts, on our infant and toddler-age experiences with parents and other care-givers.

Procedural "should" principles apply to many diverse activities: simple actions like the driving of the screw, complex interactions with others in social settings and cognitive processes like performing long division. In all these, the "how to" aspect of "should" principles blends with "is" principles.

Both "is" and "should" principles have the property that we may disregard them. We will, of course, experience the consequences if we violate valid principles. On the other hand if a principle does not direct our actions as we wish it to (either an "is" or a "should" variety), the correctness and value of that principle is called into question. In considering ethical principles, a behavioral determinist may argue that there is little or no distinction to be made between "is" and "should" principles, and that a person's envisioning of some goal and his actions in pursuit of it are only apparently voluntary, that they may, in fact, be determined by his inborn personality and abilities and by his experiences and are really inevitable. Whether or not this is true, the person feels as though he is making choices, as do groups when they act collectively to decide what ethical and political principles their members will live by and as they put those principles into words (as with the U.S. Constitution).

THE NATURES AND HISTORIES OF PRINCIPLES

Both "is" and "should" principles exist, as far as we humans are concerned, in physical, concrete forms: statements of fact or rules for behavior stored in one or more human brains by virtue of structural or chemical codes held in nerve cells, as words on a page, as pictures or diagrams or in computer discs or hard drives. These codes are formed as a result of someone's experience. Many are individual and personal. I have learned that, if I see steam rising from the contents of a pan on a stove that I must use a hot pad if I wish to pick up the pan without being burned. Experience has provided this principle. The experience of being told about principles by individuals with greater knowledge than mine also supplies me with both "is" and "should" principles. The "is" principle, that steam rising from a pan on a stove means that the pan is too hot to touch without use of a hot pad, generates the "should" principle: that I ought to use a hot pad, if I try to pick up the pan. So both direct personal experience and what we are taught lead to our acquiring and storing of both "is" and "should" principles. Both kinds of principles are taught to us by authority figures: teachers, parents, political leaders and religious figures provide us with a wealth of principles, many of which we have not learned – or which we have no way of learning – from direct experience. Reading and study also bring principles of both kinds to our attention.

It is clear that principles and rules, if they are to have any influence on what we do and how and what we think, must be both stored (remembered, recorded) and used. A principle which no one ever thinks of in planning or carrying out some activity, or which is never discussed, may as well not exist, as far as we humans are concerned. (Although, of course, scientific principles are what they are, whether we think of them or not: the earth went around the sun before there were humans, and will presumably continue to do so, whether we think about it or not).

The codes in brain and the written words expressing rules and principles have histories. They are formulated, spread from person to person, or from person to page, continue in use for varying periods of time, and change, as they are written, spoken or otherwise expressed, as the conditions in which they are used evolve with the passage of time. Scientific principles, as we perceive them, are modified by the results of new research; ethical standards change. It was once thought that one human being could own another, his slave, and could sell him if he wished as if the slave were a sack of potatoes. This attitude, and the actions it endorses, is no longer prevalent in the parts of the world we regard as civilized. A United States Supreme Court decision in 1857 affirmed the legitimacy of slavery, at least in some states (The "Dred Scott" decision). In 1865 a U.S. Constitutional amendment changed all that completely.

Principles may be lost to human memory or their records may be destroyed. The devotional practices of the Druids in Britain, and the principles on which they were based are now mostly unknown. Both "is" and "should" principles which guided the actions of medieval alchemists, as they attempted to convert base metals into gold, are mostly unknown to us.

PRINCIPLES IN CONFLICT

By their nature, scientific principles cannot conflict with each other, although they may seem to do so if the principles we have formulated, as written or spoken, are the products of incomplete or inaccurate data-gathering or if our reasoning, based on what we believe to be relevant evidence, is logically flawed. Arguments about such apparent conflicts may be vigorous and heated, but over time, collection of additional information and refining of the cognitive processes used to interpret the evidence, typically lead to resolution of the apparent contradictions and to a general consensus about the actual state of affairs. Eventually, what we say, write and think about an initially puzzling contradiction between "is" principles approaches the truth of the matter more closely. We no longer argue about whether or not the earth is flat or whether the sun moves around it or vice versa.

"Is" principles based on different methods of data acquisition and reasoning may also compete. Creationists disagree with biologists about the "is" of what people are like and how we came to be as we are. Information coming from revelation and the Bible leads those depending on those sources to different conclusions than those reached by scientists who depend on observation and experiment in their attempts to define "is" principles. Such controversies are essentially unresolvable unless one or both of the advocacy groups changes its views about what defines valid evidence or how information should be interpreted.

"Should" principles relating to ethical rules or social organization very often conflict, and these contests may go on and on. One faction in the U.S. congress holds that, as a basic principle, government should not attempt to regulate individual citizens or businesses as they carry out their activities. This faction holds that, as a trustworthy principle, since (here we have an "is" principle) people, left to their own devices, will act as if actuated by enlightened self-interest and so that the results of allowing them to do as they wish will be beneficial to the whole community. Thus we "should" refrain from enacting laws which restrict individuals' economic freedom. The dissenting faction holds that many people, if left to their own inclinations, will act selfishly, dishonestly or unwisely, with generally harmful consequences. Thus there "should" be laws regulating economic activity (our anti-trust laws reflect this view). Arguments about whether or not abortion should be legal also are about "is"-derived "should" principles.

THE FUNCTIONS OF PRINCIPLES

No bridge could be built, no meal could be cooked without use of "how-to" "should" principles derived from "is" principles. Value-related and procedural principles are involved in essentially all activity carried out in pursuit of a goal. We, and probably many animals, appear to have an inborn drive to discover rules of action which can guide what we do, and how we think. With some animals "how-to" rules appear to be inborn. "Should" rules used by a bird building a nest or by a worker bee crafting a beehive differ from "should" principles humans use in determining how to treat one another in that the bird or the bee does not evidently have any choice in how he acts. What is done can still go wrong, however if some circumstance, like the lack of nest-building twigs, prevents the proper procedure from being followed. With us, most, if not all, rules and principles, of both the "is" and "should" varieties seem to be acquired through experience. What seems to be inborn in us is not the already-formed principle, but the drive to find and formulate patterns – rules and principles – which we can use in reaching our objectives, along with the capacity to use these patterns in the process. It is difficult to think of any human activity which is not influenced by experience-derived expectations and by both "is" and "should" principles. I see a glass of water in front of me. I am thirsty, and I expect that if I drink some of the water, my thirst will be relieved. For this act to be done successfully, I must take into account both "is" principles concerning the weight of the glass and how my hand, arm and other muscles involved work, and "should" rules which apply to "how-to" organize my hand, mouth, facial and throat movements (some, but not all, of which are automatic; the "how-to" rules pertain to successfully merging the automatic and the voluntary, chosen, components of the action). My memories and some automatic motor sequences, contain patterns needed to direct drinking a glass of water, and these patterns appear in consciousness at need. Remembered principles are involved when we perform even the simplest acts as individuals or when we participate in the development of complex entities like political systems. The functioning of principles is related to how long they last in human (individual or collective) memory or in storage sites (words on a page, & so forth). Many principles, like those needed to organize the drinking of a glass of water, are reinforced by repeated use and so remain fixed in their storage sites. Other principles, especially "should" principles concerning how we must act toward others or to how we think, or to skills like the playing of a musical instrument may fall out of use over time so that they are lost to us. I can no longer remember how to play the instrument I played when I was a member of my high school band, many years ago. A librarian, who has not been asked to locate a particular reference book for several years (it is considered out of date by scholars who used to refer to it) may have trouble finding it on the library shelves; the index card which identifies and locates it may even have been lost or misplaced in the file. (Card files may have been replaced by computer programs, of course).

Changes in rules or principles may occur in several ways. Principles may be consciously and intentionally reformulated, as when amendments to the U.S. Constitution are agreed upon and recorded. At other times principles may be interpreted in different ways in different time periods, as social and political conditions evolve. Constitutional amendments are recorded and are added to the original document. Here a reader of the constitution sees something new and different when he reads it after the amendment is made. In other situations the written document may not change but its assumed meanings and implications do. The U.S. Supreme Court has at times overturned previous rulings, when no change in the written document has been made. In 1896 it affirmed the constitutionality of laws which encouraged or supported racial segregation in schools. In 1954, in the Brown versus the Board of Education case, it reversed this stand.

Some principles change through development of informal, spontaneous and unplanned group consensus. What is written in the King James Bible about Jesus' statements concerning divorce has not changed, but many (perhaps most) Christians today view divorce as being morally acceptable.

SUMMARY AND CONCLUSIONS

Principles and rules are of two general types: "is" statements about what things are like, what they are made of, how they act and change and how they have come to be; and "should" formulations which specify how it is desirable for people to act in pursuit of a goal (like driving a screw into a piece of wood or in interacting socially with each other, and in a myriad of actions having degrees of complexity between these two extremes). We encounter principles in concrete forms: as words on pages; as structural or microstructural and chemical configurations in the neurons and neuronal networks of our brains' memory (here we do not see the principles written out, but we experience knowledge of them vividly, nevertheless); as high spots and dips in the tracks of vinyl phonograph records; or as zones of varying reflectivity along the paths of compact discs. In all these sites, principles exist and persist in physically configured codes. Principles in these kinds of memory are recorded because of the ways our brains – and any artificial sensing and recording devices used for holding principles – are made, and how they process input stimuli. Nest-building of birds is probably an expression of "is" and "should" principles pre-programmed into the birds' brains. We humans appear to have few pre-formed patterns; we develop patterns mostly through experience. But we do have an intrinsic urge to find patterns representing principles and rules (including "how to" procedures).

Principles and rules function for us as tools we can use in our efforts to reach our goals. They are, in fact, essential if we are to survive. We must eat regularly (an "is" principle) and we must have strategies and procedures for finding and preparing our food ("should," "how-to" principles). Rules and principles help us to find freedom from anxiety and to provide us with physical comforts ("If I put on this sweater, I won't be cold anymore") and to experience a wide variety of satisfactions: the pleasures of companionship, feelings of accomplishment associated with winning a game or building a house, the successful exploration of an unfamiliar geographic area or of the implications of an idea, for example. Principles which do not serve some such ends and are not used regularly typically fade from individual and collective consciousness.

Conflicts of various sorts between principles have often occurred and it seems certain that this will continue to be true. History teaches us that principles change over time with social, economic and political conditions' evolution. The principle that Catholic priests should be celibate became generally accepted only after 800 A.D.; and in Ireland priests could be married – without being considered immoral – as late as 1000 A.D. That a person should "turn the other cheek" if attacked is a much newer principle than the older, reciprocal, "an eye for eye, a tooth for a tooth" rule. Considering how many of us act, it is tempting to view the older principle as an expression of an "is" principle which reflects how we are genetically programmed to respond to attack. The occurrence of family feuds in the past in Appalachia and the usual responses of a dog which has just been bitten by another dog support this notion.

It is clear that we must have rules and principles if we are to survive and thrive. It is also clear that principles have histories; they change as applied in action and they appear and disappear as our needs change over time. Principles – especially "should" principles often conflict and the conflicts may not always be resolvable to the satisfaction of all those concerned. Apparent conflicts between "is" principles generally end eventually with the growth of knowledge (a possible exception to this statement is the difference of opinion between creationists and those who believe In evolution). On the other hand, disagreements between holders of conflicting "should" principles sometimes appear destined to go on and on disagreeing.

A last thought concerning conflicts between principles. Some conflicts are potentially due to how language is used. Different people think about and express things in different ways, so that two people may in fact agree about the validity of a principle but may appear to disagree. Such conflicts can potentially be resolved with continuing discussion, "Oh, that's what you meant! I didn't realize it at first."

REFERENCES

Random House Webster's College Dictionary

Holy Bible, King James version

James, William, <u>The varieties of Religious Experience</u>, Barnes and Noble Classics, New York, copyright 1954; (see Hames' revised edition, 1902, for notes and commentary)

Roberts, J. M., <u>The History of the World</u>, Oxford University Press, New York, 1993

Kandel, E. R., Schwartz, J. H., Jessell, T. M., <u>Principles of Neural Science</u>, McGraw-Hill, Health Professions Division, New York, 4th edition, 2000

THOUGHTS ON ACTIONS' ENDINGS: EXPLORATION AND OTHER ACTIONS

INTRODUCTION

An article in a 2007 issue of the journal <u>Sleep</u> included the statement that, "In animal models (in this case, rats) exposure to novel and enriched environments enhances exploratory behavior and induces widespread plasticity in cortical and hippocampal circuits." (1) Piaget's studies of infant behavior (2) and other writings concerned with childhood development make it clear that the drive to explore novel environments or to investigate unfamiliar objects is not limited to animals in research laboratories. These sources may include discussions about what causes exploration to begin, but do not usually deal with what makes it stop. But what induces an action to be interrupted or terminated is not inherently less interesting than what caused its beginning. This set of ruminations was provoked by the idea that examining the ways exploratory actions terminate might shed light on the essential properties of exploration and some other types of behavior and on how all these actions come to their ends.

THE SCOPE OF THE WORK

Behavior structured by the intrinsic genetically determined properties of body and brain will not be evaluated in depth. The rooting and sucking actions of the new-born infant (3) and the stepping movements of the decerebrate cat (4) exemplify this kind of activity. Action patterns developed in infancy and very early childhood, like the reaching for and grasping of objects appearing in an infant's visual field will also be left unexamined. These notes are concerned with action episodes the natures of which depend, at least in part, on cognitive processes. The emphasis is on determining the qualities of these episodes and on identifying the influences which cause them to end. Specific actions – fighting, making love, playing soccer and so forth – are not included in the discussion, except as they may serve as examples of particular types of behavior. That apparently dissimilar activities may share features which allow them to be placed together in a broad class of behavior is a notion supported by the recognition that a surfer riding a wave and a person reading a book he finds entertaining – two very different pursuits – have in common that what they are doing is bringing them immediate pleasure or satisfaction; both individuals are engaged in ludic (inherently pleasurable) activity.

STOP SIGNALS

A dictionary of psychology (5) defines behavior in general as "actions, reactions and interactions in response to external or internal stimuli, including objectively observable activities, introspectively observable activities and unconscious processes." This definition of behavior implies that no behavior occurs if input stimuli do not arrive at the functional brain sites where actions is planned and set in motion. So consideration of input must be a part of any analysis of behavior, including the assessment of how episodes of behavior end. Since what we do is a response to input, then stopping an activity is also a result of input, since stopping an action is itself an activity. The knee-jerk reflex illustrates this well. When the patellar tendon is suddenly stretched by the blow of a reflex hammer, the stretch receptors in the tendon send signals to the spinal cord, where motor neurons are activated. These motor neurons send other signals down their axons to the knee extensor muscles, which then contract. The input from the stretch receptors, besides being directed to the extensor motor neurons, goes to inhibitory interneurons which have synaptic connections with motor neurons for the knee flexor muscles (4), so that these muscles are made to relax during the time the extensor muscles are contracting. The inhibitory interneurons have sent a signal to the motor neurons for the flexors, so that they stop maintaining flexor muscle tone.

Behavior of any kind must respond to a stop signal. Each individual action must stop before another can begin. In fact, every episode of behavior must end with a stop signal. Our activities are inherently intermittent. We do not continue to perform any particular activity indefinitely. It is notable that any activity in which a person is involved itself generates input. When we speak we hear the sounds we are making and feel the sensations produced in our lips, tongue and throat which result from phonation and muscular actions used to pronounce the words. We turn our heads, and the image of what is before our eyes changes. All this implies that the signals which cause us to change what we are doing at any moment are generated in response to that activity. Direct sensory feedback is part of this group of signals. Other effects of what we do, like the occurrence of mental fatigue (see later discussion) may be important as well.

The diversity of what we do argues that the stop signals which make it possible to go from one kind of action to another are also likely to be diverse. The following scenario aims to illustrate this point. John Wilson goes out to his yard to cut the grass, the first task in an afternoon he had planned to do yard work. After an hour of grass-cutting John feels fatigued, he stops to rest for a few minutes so he will have energy enough to finish the mowing. After resting for a short time, he goes back to work. Shortly after he has started, he catches his foot on a rock which forms part of the border of one of the flower beds, starts to fall, and twists his knee. As he struggles to keep his balance, he feels a sudden pain in the knee, the same one which was injured long before in a college football game, and which he has re-injured several times. He tries to keep on working, but the knee becomes more and more uncomfortable, and he recalls that when had re-injured the knee in the past, pain usually increased if he did not stop what he was doing to allow time for the effects of the new injury to subside. He abandons his plan to continue with yard work that day.

After John goes out to the yard, Sarah, his wife, picks up her knitting needles to finish the work she had started earlier on a sweater she was making for her grandson. After she is done, she starts to put away her knitting needles and yarn, but before she can finish cleaning up her work area, the doorbell rings. She stops what she is doing and goes to the door. On opening the door she sees a large carton on the doorstep and that a UPS delivery truck is just moving away from the curb in front of the house. She had not expected a delivery, but she sees her name on the box's address label. She takes the carton into the house to open it. Inside it she finds a large amount of protective packing material, and then, when this is cleared away, an object she does not at first recognize. After looking at it carefully, however, she realizes it is a lamp, of a kind she had wanted to get for the room where she does her knitting, to replace the barely usable one currently in this work area. She realizes that John has ordered the new lamp for her, knowing that she wanted one like the one just delivered. She sees and invoice which had been packed with the lamp, and feels a flash of anxiety, fearing that the lamp was unreasonably expensive. Reading the invoice, however, she is reassured to find the cost was affordable, and sense of relief replaces her earlier concern. She also feels pleased at discovering how mindful her husband had been of what she wanted. While Sarah was unpacking the new lamp, John, who had given up on his plans to continue with yard work, comes into the house and goes to his study to do something he had been putting off – planning the couple's budget. After reviewing income and expense statements for some time, he finds he is having difficulty concentrating on the figures before him. He puts the papers aside and goes to join Sarah, who has just started to prepare dinner.

After dinner the couple pick up the dishes and clean the kitchen, after which they sit down to play cribbage. After they play several games, some won by Sarah and some by John, Sarah says, "Don't you think we've played long enough? Let's quit and watch television for a while."

This story identifies seven types of stop signals. These fall into two classes. The first class is made up of fatigue, painful or disagreeable sensations or the expectation that such sensations are in prospect and the interruption of an activity by an influence not related to that activity. John felt muscular fatigue when mowing the lawn and mental fatigue while working on the couple's finances. Pain was the final reason for John's stopping yard work for the day, and the expectation that pain would increase if he tried to continue working reinforced his decision to stop. Sarah stopped putting away her knitting needles and yarn when she heard the doorbell ring. The three members of this class of stop signals might be thought of as general, universal. Any one of them could end or interrupt any kind of activity. Pain and muscular fatigue need little or no further discussion. Intrusive influences which interrupt actions in progress and mental fatigue both call for further analysis. Intrusive stop signals are not always as overt as the ringing of the doorbell which brought Sarah to her front door. If the doorbell had not rung, but if Sarah had glanced at her wrist as she was working, perhaps entirely by chance, and had seen that the time when she ordinarily began to prepare the evening meal had nearly come, she would probably have stopped what she was doing, leaving completion of her task to a later time. Sarah's daily routine matters to her; she always has it in mind.

Here, the generator of the stop signal originates in Sarah's brain and the trigger which activates it is the glance at her wrist watch.

Mental fatigue also deserves further discussion. We all experience sleepiness at the end of a day which makes it hard to concentrate on what we are doing, but even when sleepiness is not a factor, we find that maintaining attention on a mentally challenging task becomes more and more difficult as we continue with that task. A partial explanation of mental fatigue may be that neurons and neuronal networks that require rest, just as muscle cells do. Muscles and neurons need energy. Both muscle cells and neurons do something: contraction in the case of muscles cells and the generation and transmission of signals in that of neurons. Some energy must be used each time a neuron acts. The more times this activation of a neuron occurs each second the more energy is used, and the less time there is between actions for the re-accumulation of cellular energy stores, for cellular rest.. The idea that increased signaling activity leads to neuronal, or neuronal network fatigue is supported by several lines of evidence. The longer a person is awake, active and paying attention to what he is doing and thinking, the more slow-wave sleep he has in his next sleep period (6). The way long periods of active wakefulness are followed by relatively large amounts of slow wave sleep resembles the need for rest after sustained muscular work. The longer the interval of wakefulness or general mental activity (data processing, action planning and implementing of action), the longer the duration of slow wave sleep is in the next sleep episode. Slow wave sleep is necessarily the condition in which the frequency of occurrence of cerebral cortical neuronal actions is the slowest. A further indication that neurons can become fatigued and then must rest is provided the observations that, in encephalogram (EEG) recordings during sleep slow wave activity is most prominent over cortical territories which were most active during earlier performance of test tasks, as judged by functional magnetic resonance imaging (MRI) done during the tasks.

The members of the second class of stop signals which acted during John and Sarah's afternoon and evening activities differ in an important way from those in the first category. Each member of this kind of stop signal is related to the type of action it brings to an end. When John and Sarah cleared up their supper dishes, they stopped this activity because the process was finished, not because of distress, fatigue or a distracting interruption. When Sarah had removed the new lamp from its carton and had examined it long enough to discover that it was and how much it had cost, she had no further incentive to continue handling or inspecting it. She had learned all she felt she needed to know about it, like the laboratory rat which stops moving around and sniffing at a novel object just put into its cage, and begins to groom itself or goes to its food trough to eat. For both Sarah and the rat, exploration is finished.

Completion of an action must have several elements. John and Sarah's story suggests that there are at least two kinds of action completion. In the case of exploration, learning something not previously known is a necessary part of the ending of the activity. A code is formed and placed in memory. At this point the novelty is no longer a novelty; it is something familiar, which can be recognized if it is encountered later. And during exploration, codes depicting it have become associated with those already in the memory files, so it can be classified as like or unlike things already known and also as having feeling state-related properties: "pleasing," or "unimportant," or "frightening." With the formation and insertion in memory of the primary representational codes (basic data) and the establishing of associations between these codes and feeling state codes, exploration ends. The ending of John And Sarah's after-dinner clean-up shares with the ending of exploration, that it is a completion, but it differs from that of exploration in several respects. It does not involve formation of a new memory or of new associations. Instead it depends on there being already in memory a representation of how things will be when the cleaning up has been finished, and that there be – also in memory stores – a code complex which specifies the steps which need to be taken to complete the process. In this case, action completion stop signals occur because of feedback indication that all the programmed steps have been taken, coupled with the input confirming that the purpose of the job has been accomplished. The dining room and kitchen look just as they should.

Influences which end ludic activity, engaged in because of the pleasure its performance provides, have little or no completion quality. John and Sarah play cribbage because they enjoy it.

Influences similar to habituation (4) seem to lead them to put away the cards. The pleasure they feel as they play diminishes as play continues. It is as though the feeling state generating systems in their brains are sending progressively fewer reward signals as playing time is extended. A model of reduced responsiveness with stimulus repetition is what happens to the behavior of the marine slug Aplysia when it is presented with a series of tactile stimuli, delivered to its siphon (4). At first, a touch to the siphon leads to withdrawal of the siphon and gill. When the same stimulus is repeated over and over, the withdrawal movements become less and less vigorous. Eventually, after many repetitions of the stimuli, the responses stop occurring. If the slug is left alone for a time, however, it again responds to stimulation. With repeated presentations of trains of stimuli the time between trains of stimuli needed for recovery of the withdrawal response becomes longer and longer. With many repetitions of stimulus trains, the response may be lost permanently. Our responses to repeated experiences often have a similar pattern. The feeling state storage and transmission systems in our brains behave like the motor neurons of the marine slug. The sites sending pleasure signals to consciousness seem to be less and less responsive as we repeat or continue doing something we initially enjoy. John and Sarah will probably play cribbage again but they may eventually lose interest in that particular game and take up Gin rummy. We find ourselves saying, "For a long time we went almost every Friday night to that Chinese restaurant, but lately it hasn't seemed appealing and we've been going to the Seafood place by the pier instead."

Another type of feedback stop signal of those related to the activity it influences is the reaching of an envisioned goal. A college degree is in hand; Sarah finishes knitting a sweater. What a person had in mind as he started an activity has happened. There is no incentive to continue the activity. There are similarities between behavior directed at action completion and that aimed at reaching a goal, but these two activities differ in several respects. Goal directed behavior, as the term is used here, stands apart from the habitual or routine activities of daily life, and from exploration, in that it begins with the visualization of a new thing or set of conditions not part of current experience. If the envisioned goal is, or resembles, a previously accomplished one, it is not a frequently recurring objective, like seeing (and feeling pleased by) the sight of a tidy kitchen.

Sarah's ideas about a sweater as she starts to make it are different from her mental image of how the kitchen will look, after the supper dishes are cleared away. She has seen her kitchen in a state of good order many times in the past, and expects to see it in the same state again in the future, but she has not seen the sweater she plans to make. She may have knitted other sweaters in the past, but each one was the product of a separate imagine-plan-perform process.

ACTION CATEGORIES

The activities terminated by the stop signals which are particularly related to them seem to fall naturally into four general classes. Exploration and performance of programmed action sequences end when something has been finished. Exploration has caused a code complex to be placed in memory, along with its associations, which include both other data-depicting codes and feeling state codes. A habitual activity has come to the end of its predetermined series of steps. Things done for pleasure have lost some of the appeal. A goal has been reached. So there are four general types of behaviors: Exploration, performing a programmed action sequence, doing something for pleasure and working to reach a goal. Any of these types of behavior can be halted by fatigue, distress or the expectation of distress, or by an unplanned, unexpected interruption. However, even if one of the three general action terminators is not in play, an episode of any of the four types of behavior will end if input from its own particular action stopper comes to the awareness-action field.

Some further considerations concerning goal-directed behavior deserve attention. First, it should be kept in mind that exploration differs from goal-directed behavior in this important respect: exploration does have a goal, but this is general, the understanding of a new thing or place. Its goal is not consciously formulated. The impulse to explore something new is there from the start. This need to explore appears to be inborn.

Another type of goal-directed behavior to be considered here has to do with activity which is responsive to the coincidental and often unexpected events of everyday life. A man driving his car along a city street glimpses something moving in the space between two cars parked along the side of the street in front of him. He then sees that the moving object is a man walking out from between cars, that the man is using his cell phone and is apparently unaware that the driver's car is coming toward him. The driver brings his car quickly to a stop. The driver's action is clearly goal-directed, even though it has nothing to do with a previously imagined objective. The appearance of the man who starts out into the street from between the parked cars leads to the instantaneous formation of a goal, the prevention of an accident. The goal-directed quality of the driver's behavior is not made less by the facts that formation of the goal was very rapid and that it was generated by a combination of unexpected circumstances. The sudden surge of increased attentiveness and action readiness, here coupled with a flash of anxiety, is probably an inborn reaction to the sudden perception of something unexpected. These effects, while they are due to unanticipated input, do not structure the driver's responses, which are obviously determined by the details of what he driver sees and by his past experiences in responding to suddenly occurring events. What the driver sees has none of the quality of exploration, the performing of a previously established routine or of an activity pursued for pleasure.

SUMMARY AND CONCLUSIONS

There are (at least) seven types of stop signals, the arrival of which in consciousness – in our awareness-action fields – can bring an activity in progress to a halt. Three of these can terminate any kind of behavior. Aversive signals like pain, fear or the expectation of distress; muscular or neuronal fatigue; and the intrusion of influences not intrinsic to the activity being performed, but which make it impossible or undesirable for that activity to continue. The other four stop signals relate to the particular kind of behavior they affect. These are the perceptions that: a pre-programmed or habitual action sequence has come to its end; an exploration has resulted in the placement in memory of code complexes representing some novel object or situation, along with the associations formed during exploration, which relate what was explored to other memory codes and feeling state codes (for example, pleasurable feelings experienced previously when eating something which tasted good) evoked during the exploration; an habituation-like effect (boredom?) is developing; or a goal has been reached. The process of identifying the last four stop signals carries with it recognition of the distinguishing features of the four broadly definable classes of behavior related to them. It is difficult to imagine any kind of action which does not fall into one of these four classes. It is likewise hard to think of any action the ending of which is not due to one, or some combination, of the seven stop signals just discussed.

Of course an episode of behavior may have features of more than one of the four general classes of activity, but when a person talks about what he is doing, he typically gives his action a single name, that which seems to describe it best. Clearing up a dining room and a kitchen after a meal is very different from the making of a sweater. So it is with action endings. Several influences may act together to make an action stop, but one usually stands out and is named as the principle one. A person may say he now understands something he has been investigating, or that he has become bored by or has lost interest in an ordinarily enjoyable activity, or that he has finished some routine task, or that he has completed a special project.

Defining classes of any thing or action may be based on many different principles. Behavioral episodes could be identified as those which did or did not involve the use of tools; or what is done during war time as contrasted with activities characteristic of periods of peace. The classes of actions and of action terminators discussed here are very broad. This approach to classification was used so that the classes identified would include as much of the range of human (and animal) behavior as possible. This approach to classification has a major limitation in that any classifier's imagination, his concepts of how classes can be defined, is sure to be less than perfectly comprehensive. But, hopefully, the classes here employed are suitable tor this attempt to explain how episodes of behavior end. It is evident that this set of ruminations, which began with an interest in exploration, is itself an exploration.

REFERENCES

Huber, R., Tononi, G., Circelli, C. Exploratory Behavior, Cortical

BDNF Expression and Sleep Homeostasis. Sleep 2007; 30, 2:129-139

In: Piaget, J., The Origins of Intelligence in Children. New York: W. Norton & Company, Inc. 1963 (original copyright, International Universities Press, Inc., 1952

In: Caplan, F. (general editor) The First Twelve Months of Life: Your Baby's Growth Month by Month. New York, Toronto, London, Sydney, Auckland Bantam Books, Grosset and Dunlap, Inc. 1971-1973

In: Kandel, E.R., Schwarts, J.H., Jessell, T.M. Principles of Neural Science, fourth edition. New York McGraw-Hill, Health Provisions Division 2000

In: Corsini, R. The Dictionary of Psychology New York and London Brunner-Routledge (The Taylor and Francis Group), 2002

In: Kryger, M.H., Roth, T., Dement, W.C. (editors) Principles and Practice of Sleep Medicine, third edition New York, London, Philadelphia, St. Louis, Sydney, Toronto W.B. Saunders Co. 2000

US AND THEM

GRAMMAR AND SEMANTICS

"US," "THEM," "WE" and "THEY" are pronouns. "US" and "THEM" are in the objective case; "WE" and "THEY" are in the nominative. All four words are plural forms, and point to the existence of groups, of one of which, the person speaking or writing the word or words, thinks of himself as being a member. Each group is defined by its typical characteristics, or by what its members commonly believe to be these characteristics: tallness, being a male, having blue eyes or dark skin, being a Baptist or a republican. "WE" and "US" feelings clearly occur in animals, considering how herds of grazing animals and bees in beehives behave. People often use "THEY" or "THEM" to refer to non-human entities, "Be careful if you go near the nest those wasps have made – they might sting you." The discussion which follows is focused on human groups, both "US" and "THEM," however.

REGARDING GROUPS IN GENERAL

The first defining phrase for the word "group" in the <u>Random House Webster's College Dictionary</u> is, "any assemblage or cluster of things, an aggregation." Another expands on this first definition by stating that a group is, "a number of persons or things ranged together, or considered as being related in some way." A group is thus a class to which individuals may belong if they share its defining properties. Group members need not be in all ways identical, however. They may differ from each other in ways unrelated to the group's defining features. Boys and girls are both children. Group members cannot be perfectly identical, in fact. Each person has his distinctive de-oxy-ribonucleic acid (DNA) gene pattern, and individuals within many groups vary widely in appearance, size, personality and abilities (Consider a class in high school. All its members do not earn "A" grades, or join the football team, but they are still group members, "Class of 2009.").

"WE," "THEY," "US" and "THEM" refer to how a group member thinks of himself, to how he views other members of his group and to how he thinks and feels about persons who are not group members. "We're always disappointed when our team loses, but we don't raise hell and say the coach should be fired like they do." This sentence points to the fact that we see our "WE" as different from "THEM" (and to our feeling disapproval for how "THEY" act). They may also differ from us in other ways, besides the ways we respond to our team's performance: in skin color, or in believing – or not believing – in evolution.

Another aspect of the US-THEM distinction is that "WE" are less familiar with what "THEY" (the members of "THEM") are like than we are with members of "US," both as individuals and in relation to how individuals in "THEM" think and behave as a group. We know, in general, how the people we meet on the street in our home cities or neighborhoods are likely to behave – whether they are more likely to be polite than rude or aggressive, for example. When we go to a city in a foreign country, on the other hand, we feel less confident about how the people we meet are likely to behave. The country and the people who live there are unfamiliar.

Groups may exist only for brief periods or they may endure for years or centuries. "WE" may be the people who saw an accident on the street near where they were walking. In this instance, the group – defining property is, "those individuals who saw the two cars collide at the intersection of Market Street and Tenth Avenue last Tuesday." This group dissolves as the witnesses leave the accident scene, although it may be temporarily reconstituted when its members are called upon to testify in the course of a suit brought to determine the cause of the accident and which of the involved drivers was responsible for it. The "WE" of citizenship in a nation may last for many years, or even centuries, much longer than the lifetimes of those who originally made up the group or who lived through the later period of its existence. Long-lasting groups can and do change their defining characteristics over time. The Republican Party of Abraham Lincoln's time was not the same as that of the present twenty-first century; the doctrines of the Catholic Church have changed substantially since 500 A.D. (with respect to celibacy in the priesthood, for example). It may be reasonable to consider whether or not any group with substantial endurance in time can avoid changing in one or several ways as its membership changes (old members die, new ones replace them) or as its circumstances – the character of the nation of which it is a part, for instance – evolve. Further, most groups are probably not eternal: there are no longer any

Neanderthals.

FORMATION OF GROUPS AND THE DEVELOPMENT OF GROUP MEMBERSHIP IDEAS AND ATTITUDES

Identifying the ways individuals develop "WE" and "THEY" ideas and attitudes and how groups form in general is central to understanding the "US – THEM" phenomenon. Some group memberships are involuntary and inevitable: that of the people who witnessed the accident; being a citizen of a particular country; being Negro or Caucasian, for example. It is likely, or probably unavoidable, that the first beginnings of "WE – US" ideas occur in infancy. For a new mother, the feeling that baby plus me make up a "WE" or "US" probably begins at, or soon after, the infant's first cry (this feeling may not be generated by that cry, of course, with mothers who come to the ends of their pregnancies when they had not wanted to be pregnant in the first place. With the baby, the "US" must evolve more gradually, in stages. Whether "US – THEM" and "WE – THEY" ideas and feelings are produced by experience or by a programmed process which depends on the maturing of the brain might be argued. It may not matter which way the feelings and attitudes develop, however. There is a sequence of occurrences which must take place in either case. Before there can be a "WE," there must first be an "I" in the baby's mind, recognized as separate and different from "NOT-I." (Everything which can be seen, heard, felt, smelled and tasted which is not part of the self). If a moving object is seen at the same time an infant feels tactile and proprioceptive sensations originating in his hand, sooner or later the notion, "that's my hand, a part of me, that I see moving" must be formed as the visual and somato-sensory input sensations occur repeatedly together. If the infant sees moving objects and does not have such tactile or proprioceptive input at the same time, what he sees will eventually be regarded as "NOT-I." Smiling, crying and signs of restlessness must a be provoked by something a baby sees or feels, but the sensations which set these responses going do not at, and soon after birth, appear to be recognized by him as coming from either within, or outside of the self. The baby also does not appear to discriminate between sensations associated with different caregivers' actions, in the first month or two of life. It does nothing, at this early state in development, which indicates that it can distinguish "I" from "NOT-I" or that it can tell the difference between one person who picks it up and another – between its mother and a stranger. A little later, between two and three months of age it starts to react differently to different people, to cry when picked up by someone unfamiliar and to stop crying and appear to relax when the stranger hands it to its mother. Self and not-self discriminations seem to start to be made in this same stage of development. It becomes possible to tell, eventually, from how the baby behaves, that us-them ideas, images and feelings are becoming more and more clear-cut. Over the early months of life, experiences like classical conditioning must occur. The newborn and very young infant can have no images representing how his parents look – or indeed any person or thing – implanted in his memory. He can have no way to tell one person from another or of forming the expectation that, if he sees his mother or hears her voice, he is likely to be picked up and cuddled (although, if he is in distress and crying, he may quiet and appear to relax if he is held firmly but gently by his mother). Expectations develop over time with repeated experiences, and, sometime after he is a month old, the baby begins to react differently to the sight, touch, or voice of one person as compared to another. These response differences are not well-developed at first, in early infancy, and before "I-NOT I" distinctions have become well-established. With the appearance of the ability to tell people apart comes the capacity to distinguish familiar from unfamiliar objects and people.

It is inevitable that, in his early life, a baby will have often-repeated experiences with how others respond to his needs and actions and how they act in different circumstances. Once he can distinguish self from non-self and how the individuals in his world differ from each other, and can begin to tell that special situations have characteristic features – how morning is different from evening, how his bedroom is distinguishable from the family's living room (some things happen in one place, other things in another) – he can begin to form associations involving his observations and actions, and how, for instance, he can expect his caregivers will behave at any given time. The expectations which develop provide the basis for "If this happens, then that will follow" ideas, including the sense of how the members of his "US" are likely to behave in relation to him and to each other in different circumstances. The infant learns what kinds of behavior lead to wanted, agreeable experiences and sensations and what other kinds are likely to be followed by his (or others') distress or rejection (mother speaks sharply and puts him down in his crib when he is feeling a desire to be held and soothed). A conservative theologian might argue this point, but is clear that from early infancy onward, the baby will have experiences which promote his developing behavioral standards (or rules). These standards might be viewed as the first beginnings of his sense of what is ethical or moral, and that they also form part of the defining behavioral rules of his family, his first "US." Information of this nature must enter and be retained in a baby's mind as part of what Jean Piaget referred to as "memory in the wider sense," some of which is acquired during what he called the "sensor-motor" stage of development, before event memories begin to form. "The first thing I can remember" is an image or occurrence of the toddler years for most people – how the bedroom looked or how it felt that time when the individual stumbled and fell after miscalculating the height of a step outside his front door when he was being taken along with his parents on a visit to friends.

MORE ON HOW GROUPS FORM

The sense of group membership grows and is structured more and more as a person matures. Early in the primary school years it is still mostly imposed, rather than being developed consciously. A kindergarten student has not chosen to be a member of his class. Membership, and a person's sense of membership, in social or economic classes are also generally involuntary, as is membership in a racial group. A Negro or Oriental child may not think of himself as a member of his ethnic group at first, but he comes to have this idea, once he sees that there are others in his world whose skin color differs from his, or whose facial contours do not resemble his or those of members of his family (who are at first the only people with whom he has any experience).

Group membership can be determined by coincidence (as in the case of witnesses to a traffic accident) and by unplanned events. A prisoner has become a new member of the prison population after his conviction and sentencing for a crime. A worker joins a union when he is hired by a company having a contract with the union which requires that all employees at the plant or business be union members. What causes an individual to join a group (and, in fact, what makes the group form in the first place) may be a matter of conscious choice, "wouldn't it be great if we could start a chapter of the Elks here?"

PERSISTENCE OR DISSOLUTION OF GROUPS AND NOTES ON HOW GROUP MEMBERS CONTINUE OR END THEIR MEMBERSHIPS

The existence of a group can end, when the reason it was formed no longer has force. The individuals who made up the exploring team of the Lewis and Clark expedition in 1804 separated in 1806 when the team returned to its base. This example points to several essential qualities of groups. The team formed in pursuit of a goal, unlike groups defined by the race or ethnicity of their members. During the expedition, the activities of the individuals in the team were directed both at reaching its goal and at holding the group together, at maintaining its continued existence and its ability to continue pursuing its objectives. These types of activities are not identical, but they are inseparable: it the group does not continue to exist, its goal or goals remain unmet. Some groups have endured by modifying their objectives, when the influences which led to their initial formation changed, as though their members had found that just being together in the group was important to them. The March of Dimes fund-raising organization, originally dedicated to supporting research and clinical activities aimed at eliminating poliomyelitis, still exists today even though the disease has been largely controlled by widespread vaccination since the 1950s. When its original goal was met, it shifted its efforts to combatting other health problems. The forces which make people (and animals) cluster together in advocacy organizations, flocks, herds, political parties and religious groups are clearly very powerful and diverse. Togetherness (a major member of this group of forces) seems to matter greatly to most of us.

Some relatively long-lasting groups end naturally. A student graduates from high school, and he and the others in his class are no longer members of a group with shared interests and activities. Other groups may dissolve or change if some members find that the appeal of membership has diminished or that being a member has become disagreeable. "I just hate to wear those silly uniforms at meetings." The departure from a group of some of its members may either strengthen it or weaken it. If only a small number of its members leave, those who remain are likely to be more alike in attitudes, beliefs and behavior than before the disaffected members of the group – who made the group more diverse – left. The group's cohesion and sense of consensus would thus be increased. On the other hand, if many members are dissatisfied, the unity of the group is likely to be weakened, so that the group would dissolve or change its character in ways that would increase the general appeal of membership. A majority of group members may expel a non-conforming individual if he is seen to be "not really one of us."

Some groups form, when a number of individuals come to agree that a particular goal is worth pursuing, as with the March of Dimes organization, or the Lewis and Clark expedition. During this pursuit, and indeed with most groups, members are expected to follow, and conform to, a set or a number of sets of rules and principles. These might be thought of as the ethical or moral principles of the group. This notion – itself, in a sense, a rule – is explicit for citizens of a nation (who must not break the laws) or the members of a religious denomination. It is sometimes assumed or taken for granted, even if not formally included in a group's statements of its standards (written laws, pronouncements issued – and published – by religious leaders like the Pope). "We just don't do things like that." A group's body of rules and principles fall into two broad classes: what relates to how group members think and act with respect to each other; and what they take as appropriate (perhaps obligatory) in thinking about, or interacting with, "THEM," persons outside the group. Within a group people are expected to act in ways which maintain its special properties and continued existence, as when church members gather to worship by participating in the rituals and routines of the church services (singing hymns, and so forth). Group members are ordinarily expected (or obliged) to further the survival and well-being of other members. Members of "US" are also expected to act so as to help the group reach its overall goals. In most (all?) groups, members refrain from acting aggressively toward each other and are expected to help each other in pursuit of their personal, individual objectives when they can, including participating in mutually pleasurable activities.

The rules and standards, which are parts of the things which define the group, including how the members of "US" communicate, cooperate and help each other, are, for all practical purposes, the laws, ethics and moral principles of the group. Some standards are developed consciously and formally (the U.S. constitution); some evolve more or less spontaneously. We are led back to considering again the experiences of infancy and childhood. These include, besides their producing "I" and "WE" perceptions, experiences with the potential for generating "RIGHT" and "WRONG" beliefs and attitudes. These learning experiences lead to the development of expectations. Particular conditions in which a baby finds himself, and particular things he does, give rise to agreeable sensations: relief of hunger; being moved from an uncomfortable position, or one which has become uncomfortable, to a better one (after all, the very young infant cannot turn himself over); the change from being too hot or too cold to being pleasantly warm; sensing comfortable contact with something or someone (the month-old infant stops crying when picked up and held). Other conditions or actions come to be associated with the expectation of the occurrence, or the continuing of unwelcome, distressful feelings. So pleasant sensations come to be seen as related to "right" behavior and unpleasant ones to "wrong" actions.

How group members are expected to think and act with respect to each other is often in sharp contrast to how they relate to "THEM." A soldier tries to help the members of his squad during a battle in every way possible, while at the same time he attempts to kill any enemy soldier he can find. The actions of individuals in one species toward each other can be very different from those directed at members of other species. Cannibalism is strongly discouraged, but we feel no reluctance to eat pork chops. We trap and kill rats and mice when we find they have come into our homes.

"WE," the members of "US" must have much in common or we would not have assembled into a group in the first place, but "WE" cannot be identical. It is, in fact, desirable that there should be some diversity within "US" if the group is to survive and thrive. "WE" vary (inevitably) in strength, talent and skills. Priests, nuns and parishioners are all parts of the Roman Catholic "WE," but each of these sub-groups plays a different – and essential – role in the activities of their church. An army made up of common soldiers, petty officers, line officers and generals cannot function if each of the these different sub-classes does not do its part in the fighting of a battle.

There are additional features of the fact that behavior involving interactions between people in "US" has one set of standards and behavior directed at "THEM" has another, different, one. Hostile or destructive actions directed at them seem often to be justified by the authority of group-defining principles or belief systems. The war between Iran and Iraq, often referred to as a Shiite-Sunni conflict and the Crusades of the Middle-Ages come to mind in this context.

ADVANTAGES OF GROUP MEMBERSHIP

There are many things which can be done successfully only by groups. A savage searching alone for edible roots or berries is less likely to find them than is a group of individuals searching a patch of forest as a team. People (or animals) huddle together for warmth in cold environments. They may combine and cooperate in fighting off the attacks of other groups seeking to take away their food, living spaces or sources of energy. People in a rioting mob may evade criticism or punishment for performing violent acts they, as single individuals, would never dare to do (or choose to do) alone.

Our daily lives bring us repeated reminders of group membership's benefits, many of which are due to the diversity of different group members' talents and activities. We would not have roads, potable water or reasonable safety in our streets were it not for the actions of a number of sub-groups within "US." Clearly our individual survival, comfort and convenience depend in large part on our membership in "US."

Individual's goals are more easily reached when those persons are members of groups than would be the case if they had to accomplish everything they needed or wanted to do in isolation. Finding and maintaining access to food and shelter, in most environments, calls for the expenditure of enough time and energy so that there would be little left over for learning new skills, devising tools, exploring unfamiliar terrain or painting pictures (even on the walls of caves). When we live in groups a positive feedback effect operates, so that both the individual and the group of which he is a member benefit. The person (or persons) who invented the screw driver made many of his own tasks easier (or possible) and at the same time provided the others in his "US" with a new tool.

EMOTIONAL DIMENSIONS OF GROUP MEMBERSHIP

The feelings associated with belonging, of being bonded to others like ourselves, is a major benefit of group membership. A person alone seeks companionship, birds assemble in flocks and solitary confinement is seen as a more severe form of punishment than being confined in prison with or near other prisoners. There are reasons to believe that this tendency to seek the company of people we like and with whom we feel comfortable begins in infancy. Babies less than three months old have been seen to start or stop crying according to who is holding them and to act disturbed when mother leaves them, though they do not cry if an unfamiliar person disappears from their fields of awareness (Caplan). A baby's different responses to different people argues that he is beginning to develop associations and expectations concerning how the presence of well-known people, in contrast to that of unfamiliar persons, will affect him. Experiencing pleasure or a sense of security related to being with familiar caregivers begins to be expected, even though the parents may at times do things he does not like; most of the time, mother's being there is pleasant or reassuring.

There are reasons to believe that infants' early experiences with caregivers can affect their abilities to bond with others. Babies cared for in institutional settings, in which they may be cared for by one individual on one day and by a different person on another, and in which they have fewer and shorter experiences with being held and cuddled than is usual in a family environment, appear often to fail to develop emotionally in the way that infants living in families do. (Kandel, et al.). They may have delayed development of language skills. Monkeys isolated during the first six months of life have been found to be impaired in their ability to interact socially with other monkeys. These observations suggest that abnormal social experiences in infancy can have an effect on an infant's development of the sense of belonging to an "US." "WE" and "US" attitudes and expectations, which have the potential to form when a baby's interactions with caregivers regularly involve the same individuals (family members or family member surrogates), are associated with the predictability (from the infant's perspective) of what will happen with the sight of, the touch of or hearing he voice of someone well known. The ability to differentiate familiar and unfamiliar persons develops after the first one or two months of life. Our screening of input into the categories "familiar" and "unfamiliar," once it has first developed (it does not appear to occur in newborns) clearly continues our whole lives, and is central to how we react to our ongoing experiences.

Judging by how we, and many animals, behave, the perceptions that we are encountering something unfamiliar provokes two (at least) potentially competing or inconsistent urges, both of which are essentially exploratory. The newly met entity or situation may call for investigation because of its possibly offering some pleasure or satisfaction (becoming familiar with initially unfamiliar things in the past has sometimes been agreeable); on the other hand, some things which were new to the individual have turned out to be sources, or to be associated with, pain or discomfort. Something new may be expected to be connected with either pleasure or pain. A dog strains at its leash as though it is trying to come close to something it sees or smells beside the path it and its master are following. A toddler leaves his mother where she is pushing a shopping cart along a supermarket aisle and reaches out to grasp something on one of the store's shelves: "No, no, Johnny, don't pick that up, we don't want to buy it!" A rabbit, seeing a hiker approach, runs away, as though the sight of the unfamiliar hiker has frightened it. Thus, we value predictability in our encounters with new persons or situations, and at the same time, are attracted to novelty, especially if we have had rewarding experiences in the past when investigating something new.

Considering of how we react to familiar and unfamiliar people and situations leads to the notion that there are feel-good rewards, welcome emotions associated with being "one of us," being with others who think as we do about the group to which we belong: why it exists and how group members should act (and think). Besides the consensus within an "US" about how its members should act, group members typically agree about the group's beliefs and traditions. Methodists have different belief systems than Roman Catholics. Group members (historically) have gone to great lengths in supporting the establishing and continuing existence of their theological positions. Consider the Latter Day Saints practice of young people's devoting time and energy to missionary activities, based on the assumption that their beliefs were in important ways more "right" than those of members of "THEY" they went to convert. Fans of athletic teams have some of this same attitude, as do patriots. Promoting the interests of "US" is worth struggling for, even, in some circumstances, worth dying for (consider religious martyrs). Abandoning a person's beliefs and principles (which came with group membership) has been, for many, worse than death. Continuing to support, and to act in support of, those beliefs has clearly had powerful emotional rewards. A member of "US" who acts to promote the group's beliefs is clearly involved in a more widespread activity aimed at maintaining the existence of "US," the general welfare of its members and his individual feelings of belonging, of being bonded to other group members. Thus attacks on the group or on one of its members can be seen as (have very often been seen as) threats to the security and well-being of both members and the group as a whole and of individuals making it up. Attacks of this kind are often vigorously resisted (he who disturbs a beehive is likely to be stung; a suicide bomber is usually acting - in his mind, at least – to combat attacks on his "US").

Being a group member is not always agreeable or satisfying. Persons serving sentences in prison generally would rather not be members of that particular "US," as is the case with people starving during a famine or workers unemployed during an economic depression. In such settings individuals have not chosen to be group members, but have had membership imposed on them by circumstance, or by people in positions of power or authority.

SUMMARY, CONCLUSIONS AND LAST THOUGHTS

None of us can avoid being a member of some "US." We often belong simultaneously to several groups: family, citizen, republican or democrat, plumber, housewife. Some memberships are involuntary or forced, like belonging to an ethnic group. Some are more or less voluntary (Individuals choose to join fraternal organizations like the Elks or the Rotary clubs). The members of a group, except in the cases of evanescent groups, consciously or unconsciously accept the beliefs and the standards of behavior which are parts of the defining properties of that group. What causes membership in a

"WE" may not be the same with one group as with another. How I interact with my dog is not like what I do in relation to a co-worker. The sense of unity, of members having like interests, ideas, beliefs and goals, is likely to be stronger in small groups than in large ones. I am a United States citizen, but I have relatively little sense of kinship with citizens who live in other states or who belong to a political party of which I disapprove. However, a nation is more likely to endure as an entity, and to have citizens who are reasonably safe and satisfied, if those citizens see themselves as having important interests and concerns in common. This principle applies to supra-national groups as well as to countries. The fact that so far there has not been a nuclear war argues that world political and military leaders have had enough of the conviction that "WE" and "US" include the citizens of all nations to restrain any impulses they may have had to start such a conflict, besides their realization that they themselves would be likely to suffer the harmful effects of such a war.

Considering that expectations arise out of experience and that the impulses they induce to act vary according to whether the situation encountered is familiar or unfamiliar, it is proper to include the notions of familiarity and unfamiliarity in an analysis of the us-them dichotomy. We, and many animals are clearly programmed to respond differently to things or people that are seen as well-known and familiar, as compared with those perceived as novel or unfamiliar. When one of us meets a friend, we have some idea of what to expect. On the other hand, when a person finds himself in a novel situation or confronted with a person never previously met, we feel (at least) two different, sometimes incompatible, urges. I may choose not to order an item listed on a menu in a restaurant if I do not recognize it (one kind of avoidance reaction), asking instead for something I recognize because of having had it in the past and having enjoyed it. On the other hand, I may order something unfamiliar in hopes of discovering a new source of satisfaction.

And finally: Belonging matters, feelings of being a member of a group contribute to a person's sense of security, to his pleasure and satisfaction as he pursues his ongoing activities and to his expectations concerning the future. It also contributes to his sense of self, to what kind of a person he believes himself to be, to what actions he should undertake (or avoid) and to how he should think. Our "US" memberships define us for ourselves and for others. We are in many ways incomplete if each of us is only an "I."

REFERENCES

The widely shared experiences of everyday life.
News media, especially The Seattle Times daily paper, <u>The New Yorker</u> magazine, and the local NBC affiliate and the Public Broadcasting Company.

Caplan, F., <u>The First Twelve Months of Life; Your Baby's Growth, Month by Month</u>, Bantam Books, 4. New York. Originally published as 12 booklets in "Book Binder" or 'Book Box" in 1972, Bantam edition 1978, Grosset and Dunlap, the Princeton University Center for Infancy and Early Childhood

<u>Random House Webster's College Dictionary</u>, Random House, New York 1991

<u>The Dictionary of Psychology</u>, Corsini, Raymond J., editor. Brunner-Routledge, New York, 2002

<u>History of the World</u>, Roberts, JM, Oxford University Press, New York, 1993

<u>Encyclopedia of World History</u>, Laner, William J., editor, Houghton-Mifflin Co., Boston, 1948

<u>The Anatomy of Melancholy</u>, Burton, Robert, (1577-1640)

Emorey, K., Allen, J.S., Boruss, J., Schenker, N., Damasio, H., <u>A Morphametric Analysis of Auditory Brain Regions in Congenitally Deaf Adults</u>, PNAS 100, No 17: pp 1004-10054, August 19, 2003

<u>Diagnostic and Statistical Manual of Mental Disorders</u>, Fourth Edition (DSM IV – TM) American Psychiatric Association, Washington, D.C. 1994

<u>Principles of Neural Science</u>, Fourth Edition, E.R. Kandel, J.A. Schwarts, T.M. Jessell, editors, McGraw-Hill, Health Professions Division, New York, 2000

<u>Memory and Intelligence</u>, Piaget, J., Inhelder, B., Basic Books, Inc., New York, 1973 (First published in French by Presses Universitaires de France, 1968)

CONCERNING SOME ASPECTS OF THE ASSOCIATIONS IN MIND BETWEEN SENSATIONS, IDEAS, EMOTIONS AND MEMORIES

INTRODUCTION

The idea that exploring the nature of associations could be rewarding was brought to mind by the realization that several kinds of subjective experiences commonly taken for granted as ordinary and uninteresting, are in fact complex and imperfectly understood. One of these is the unexpected appearance in consciousness, in the field of attention, of an idea or image which is not related to the ongoing happenings of the moment. "That tune just came into my head, and I don't know why." Or "when I was waking up Saturday, I was thinking of the word 'promulgate,' and realized that I didn't know its exact meaning." (Quote from a recent conversation.)

Another, probably related, experience of the coming to mind of something not obviously related to the events of the current moment is the unexpected feeling of an emotion apparently not connected to the 'here and now' occurrences, "What George's wife said to me at the party last night, and the way she said it just made me mad." When a melody or a word comes unexpectedly to mind it may be that reflection can identify a chain of association elements which connect the woman's comments with the listener's associations, "Oh, I know! We heard that melody at the concert we went to last week, and I just saw a woman who looked like the person who sang it there." But this is not always the case. It is clear that some (many?) things go on in our minds which are not the result of conscious intent and which take place below the level of full awareness.

This study rests on the assumption that associations are matters of the brain – how it is formed and how it works. The facts and concepts made use of in these notes are those about which there is general agreement in scientific circles and in the wider community with respect to their validity. Matters about which there is continuing controversy are left out. Whether or not there exist spiritual or supernatural things, beings or influences are issues and questions outside the scope of this work.

The sources of information employed in developing the ideas presented in this set of ruminations are: standard biomedical textbooks; general and specialized dictionaries; journal articles on psychology, neurophysiology and information theory; works concerned with childhood cognitive development; biographical writings; and the widely experienced events of everyday life.
The range of subjects which might be studied in relation to the idea of associations is broad. The Random House Webster's College Dictionary (1991) lists seven meanings for this word, all of which are individual, but all of which share the notion of connectedness between or among people, ideas or things. This work focuses on the entities referred to in its title which come to our attention over and over, the appearance of which affects how we perceive our surroundings and ourselves, and how we respond to what happens to us and around us.

ASSOCIATION EVENTS

The only way a person can know that his brain contains an association is if one or more of its elements appears in his consciousness, his field of awareness. Some image, idea or emotion not initially present comes to mind and remains there for a time, until it is replaced by some other attention field input. The types of associations with which this discussion is concerned – images, event memories, emotions and so forth – are in memory (or in storage in some other functional domain, as with emotions) until some stimulus brings them to awareness. This stimulus-response occurrence and the continuing presence of the association element or elements in awareness while they are involved with the individual's cognitive activity is an association event.

The input stimulus which triggers the association event is not itself part of the association evoked. It may be part of an ongoing reasoning or calculation activity or it may be an unanticipated, coincidental experience, as when a particular word or mention of an event is part of a casual conversation. It may or may not be recognizable as the cause of the association event, as was pointed out at the beginning of these notes.

A special kind of association event is the arrival in awareness of something which was part of a dream or of a word, idea or image which is unaccountably in mind on first awakening from sleep. Like association events occurring in full wakefulness, it may or may not be understandable by consideration of recent events, preoccupations or experiences. The appearance of association elements in this setting makes it clear that an association event can take place as a result of the activation of element storage sites due to information processing or manipulation taking place entirely in the brain which, while it may relate to events of the previous day or to ongoing projects with which a person is concerned, is not made to occur by the sensory input of the moment. It is notable that the activation of association elements in storage, or of the feeling state elements linked to them, occurs automatically, not because of having been willed consciously (though associations can be brought to mind on purpose, of course, as mentioned elsewhere in this study). This association element activation must be occurring constantly in waking life, as the experiences described in the introduction to this work illustrate.

ASSOCIATION ELEMENTS

Association elements – visual images, word sounds, touch sensations, emotions, memories for facts and events, and so forth – cannot exist in consciousness, in the attention field, as pictures or tiny models. There is not a miniature dog in the brain of a person who sees a dog. Rather there must be a symbol or code representing a dog there, individual and distinctive, capable of going to memory and of being brought back to the attention field at need. Coded information moving from one functional domain to another, as from a primary sensor like the retina to the attention field, must travel along pathways made up of neurons. The only thing a neuron can do is to "fire," to undergo a depolarization – repolarization process. A firing (or "spiking") event begins when a receptor on the cell body or a dendrite of a neuron is stimulated by the arrival of a chemical neurotransmitter or of an electrical impulse. Neurotransmitter molecules may be released near a receptor by the end of the axon of another neuron or they may come to the receptor via the blood stream. Some drugs, for example narcotics, may act as neurotransmitters by combining with receptors. Activation of a receptor site is followed by a change in the surface of the cell near the receptor. In the region in which this change occurs, the difference in electrochemical potential between the outer surface of the cell membrane and the interior of the cell, which difference is maintained in the resting state of the cell, is lost. A zone of depolarization appears, and then spreads through the cell and out its axon. Repolarization follows the moving wave of depolarization; the whole neuron is not depolarized at once in any single spiking. A new depolarization wave may be starting to form by a receptor before an earlier one has reached the end of the cell's axon. Transmission of depolarization waves through neurons amounts to the movement of a pulse of energy like the travel of a pulse of electrical energy along a telephone line (or of a unit of light energy through a fiber optic strand). Individual pulses are all alike. Information transfer depends on the sequencing of pulses into distinctive clusters or codes. The direction of code movement, in a telephone line or in a brain, is determined by where the energy pulse enters the wire or neural pathway and by where the wire or fiber optic strand, or the pathway made up of neurons, starts and ends. In both cases, the energy pulses can go in only one direction. With the telephone, energy movement starts at a specific point and goes away from that site; in the brain, the nature of neurons dictates that signal codes can go only from axon to target cell receptor and from target cell receptor to a secondary target cell, and so on.

The foregoing discussion leads to the conclusion that association elements must be binary codes or code assemblies. These digital code units, besides being transmissible along neural pathways, have to be capable of being stored in an inactive state and of being released from storage in order to be sent to the attention field or to other functional domains, perhaps to other memory repositories or to feeling state holding and projecting loci. The release and transmission of code units from the domains in which they rest essentially inactive has the quality of a response to a stimulus. The mechanisms by which associations are stored in long-term or in short-term and working memory, and by which they are held in the attention field during an association event and during the cognitive process of which that event is a part must be complex. Consideration of all such mechanisms is outside the scope of this set of notes (see addenda). That there are neural processes which produce these conditions and effects is certain however. We experience the results of the operation of these influences repeatedly in our daily lives.

The recognition that association elements are embodied in code units directs our attention to a consideration of what kinds of things these code units represent. Another way of approaching this subject is to ask what building blocks make up code unit schemata in general. Initial code formation for memories of things, facts, persons, and events occurs in the retinas, the inner ears, and in the ends of the sensory nerves serving the skin, the musculoskeletal system and the viscera. Each type of sensor is specialized; it responds only to the stimuli to which it is dedicated. Retinal sensors do not form and transmit signals representing sounds, or touch sensations.

Feeling state association elements, emotions, are a special class of code unit. Fear, joy and anger receptors do not occupy the skin, the ears or the eyes. Code units for emotions must be sent to the attention field from neural domains within the brain which transmit their characteristic signals in response to stimuli they receive from other neuronal systems or networks. Both our daily experiences and logical considerations suggest that the stimuli which produce responses in feeling state holding and sending domains come from at least two sources: sensory systems and the attention field. Anxiety is felt when a person's breathing is interrupted by unexpected immersion in water and pain experience necessarily is related to fear that the pain may continue. An example of an emotional state induced by thought is the anger felt when a person hears a companion mention the name of someone the hearer has been injured by or by whom he has been attacked in some way.

The emotion holding and sending domains are ordinarily latent. All of the distinctive, widely recognized feeling states are not consciously perceived all the time. Joy, anger, and contentment are not felt unless the loci capable of transmitting these feelings change the intensity of their output (up or down: pleasure may lead to an increased sense of well-being: the relief of anxiety, with reduced signaling to the attention field from anxiety related domains, results in a reduction in apprehensiveness and an increase in well-being). The increase in well-being which follows relief of anxiety is an example of the effect of the change in output of signals from one hold-and-send feeling state domain acting as a stimulus to another, different, feeling state domain. The emotions in this situation are blended or are presented to the attention field in a rapid, characteristic sequence. One feeling state has provoked the appearance of another. Emotion related centers may, when not powerfully stimulated, send signals periodically or at some baseline level most of the time (see Kandel, et al, regarding the effects of axotomy. Most – probably all - neurons must fire at some minimal frequency or they atrophy and may die. But we are able to feel fear or happiness our whole lives unless critical brain damage occurs). We only become aware of a particular emotion when a change in its intensity captures our attention. And while we perceive mixtures of emotions at times (joy and excitement, for example) we are ordinarily able to identify the predominant feeling state we are experiencing.

Feeling state holding and sending domains clearly have at least two signal targets: consciousness and neuro-endocrine control centers. Fear and anger lead to increased pulse rate, oxygen consumption and sweating. Contentment is associated with reduced muscle tension and at times reduced vigilance (which last effect reinforces the idea that emotional-related domains send signals to each other). It is notable that many of the effects of the actions of neuro-endocrine control regions activate primary sensory input channels. When we perspire we feel the moisture on the skin and see the shine of the accumulated sweat on our hands. We may be aware of feeling breathless when we are afraid. Often, we recognize the association between a feeling state and a set of physical sensations.

The different feeling state holding and sending domains have distinctive properties. Fear is not like joy and anger is not like sadness. The code units sent from each such domain are individual and characteristic, and in this respect, feeling state – related sites are like sensory nerves. Each site sends its specialized type of code unit, just as each type of sensory system responds to and transmits signals for which it is specialized (skin sensors, acoustic nerves, and so forth). Several lines of evidence argue for the existence and function of these systems or domains.

A rat, in which an electrode has been placed in a particular forebrain location (see Kandel, et al) and which has discovered that, by stepping on a lever, it can activate the electrode (of course it does not know anything about electrical stimulation theory, but it experiences the effects of causing the electrode to become active), will step on the lever over and over, disregarding opportunities to eat and abandoning its usual grooming and exploratory activities. This rat displays no signs of fear or anger; it does not freeze or attempt to escape from the region of the lever. The drug-seeking behavior of a narcotic addict is in many ways similar to the actions of the laboratory rat. The arrival of the addict's drug of choice at special neuronal receptors is followed by his experiencing feelings of reduced anxiety, relaxation and increased well-being. He does not appear angry or agitated.

Further evidence for the existence and function of feeling state hold-and-send domains comes from observations of subjects who have sustained damage to special temporal lobe structures. In these individuals experiences which would ordinarily cause them to feel fearful and to show signs of anxiety (tremor, sweating, restless moving about and so forth) fails to do so. It is as if a fear-transmitting center or pathway has been destroyed or isolated so it can no longer send signals to the attention field.

The conclusions which emerge at this point are that associations (and what connects their elements) are code units and code unit assemblies, held in storage in the brain from the standpoint of the attention field between association events, and that these schemata are of two types. One of these is memory for facts, events, and concepts. This class of schemata encompasses codes generated by experience and structured by assembling and combining codes derived from the several sensory systems (vision, tactile sensation, proprioceptive feelings and so forth). The other is related to emotions. The centers or networks from which association schemata are sent to the attention field and their other destinations are specialized. How these functional domains are formed and how they function is determined by the inborn properties of the brain. What is held in the memory class of the association element storage domains is experience – derived, ultimately structured, out of sensory system code unit signals. What is retained in the feeling state hold-and-send domains is essentially genetically determined and inherent in the properties of the brain.

INPUT SCREENING

An association event is a response or series of responses to an input stimulus The word input, in these notes, is meant to refer to a code unit signal which, when it comes to its target causes some action to take place. Input may be the result of stimuli received by one or several of the various sensory systems (vision, hearing and so forth). Much, or most, of such input is directed to the attention field. The attention field may also be the target of signals originating within the brain as with dreaming; of memory, as when code units appear in consciousness in ways like those cited in the introduction to these notes; and of functional sub-domains within the attention field itself and short term and working memory systems. Input signals may be directed to a variety of targets, among which are single neurons, physiologic systems regulating cardiac and respiratory function and memory storage bins, in addition to the attention field (see later notes regarding input and the actions it may induce).

What results from input in relation to association events is not random. The sight of a friend, the sound of a word or the idea which surfaces in consciousness at the end of a reasoning activity are all processed before they provoke or become participants in the structuring of a response. The result of synaptic input into a single neuron is a special case. In this instance, only one response can occur: the firing, the depolarization-redepolarization of the target neuron. Each firing of the neuron is like every other one. But even here there is some potential for response variability, in that the neuron may fire at different frequencies or with different spacing of firing clusters at different times. Input coming to all other, more complex, functional domains, may lead to more diverse responses. It is also important to recognize that, for many input targets, and especially for the attention field and the various types of memory bins, the volume and diversity of input signals is very great. We cannot take note of or think about all of what we see and hear: we can place in memory only a part of what our sensory systems provide to us or what our reasoning processes produce. A person does not consciously think about the sensations of his feet striking the ground as he walks. These considerations point to the conclusion that input screening must be in operation in our brains essentially all the time. This screening determines both whether input signals reach the targets at which they are potentially directed and which, of several possible targets, they affect. The blocking, during sleep, of input which in waking life would ordinarily come to attention, when thalamic gates close and prevent sensory signals from reaching consciousness, is an example of one of the consequences of input screening. Another is the prevention of sensations produced in the feet while a person is walking from reaching intermediate or long-term memory, although these sensations remain in ultra-sort term memory as each stepping movement is completed. A student, working in a noisy or potentially distracting environment may block input unrelated to the subject of the book he is reading. He does not take note of what is going on around him, nor does he later remember what was occurring in the place where he was studying, beyond the recollection that his study venue was potentially distracting. It is clear that only part of the flow of input signals appears in the attention field or goes to intermediate and long-term memory and that input screening is what determines what happens to input code units. Screening and signal directing activities do not affect signals arising only in sensory systems. Daily experience repeatedly shows us that attention field input coming from memory can be blocked or can have its period of residence in consciousness shortened (any association element which appears in the course of an association event must remain in consciousness more than momentarily, if it is to be recognized, and used in a cognitive process). "You shouldn't have said that! Thinking about it just makes me angry." The speaker then starts to talk of some less disturbing subject. The association event experienced by the complainer involved both content elements, like images, facts or memories of past events, and feeling state elements linked to that content. For the duration of the association event some influence, necessarily dependent on signal movement from neuron to neuron, or on chains or loops of neurons linked together, served to maintain the elements in consciousness. A screening action, in this case conscious and purposeful, blocked this element–sustaining code unit travel and allowed other material to enter the attention field. There is good reason to believe that input (to the attention field) blocking of this kind can be largely unconscious. In the post-traumatic stress disorder (PTSD), minor events act as association event triggers, bringing to awareness painful memories or provoking a startle. Individuals who have had experiences like those of subjects with PTSD, but who have not suffered from this problem, do not have the intrusion into awareness of the content elements and emotions with which provoke intense feelings in those with the PTSD.

Thus, PTSD might be viewed as, in part, a failure of the screening and blocking processes which ordinarily regulate attention field input and send to erasure (or allow to fade from awareness) input signals classified as not likely to be associated with negative feeling states' input into the attention field.

ASSOCIATION EVENT TRIGGERING

The triggering input for a particular association event cannot be part of the association it activates, since the association elements and what connects them are already in storage in memory and in the feeling state hold-and-send domains before the event occurs. On the other hand, triggering signals must be generated by sensory systems or by the results of current cognitive activities. They are code units of "now" rather than being items in storage. Despite this difference between "now" and "then," there must be a link of some kind between the triggers and the associations they activate, perhaps one related to the kinds of factors binding the associations' elements together. A trigger can duplicate something in memory; a person sees a reproduction of a picture he has seen previously. This brings to mind the museum in which he saw it first and how that earlier experience made him feel – interested, disgusted, delighted.

For other association events the trigger could have other kinds of connections with stored associations. One type of connection could be (almost certainly is) based on similarity between the trigger and an association element. The object or person just seen resembles or has properties like those of an element stored in memory. As I walk past the workbench in my work room, in which, besides the workbench and tool rack, there is a set of shelves on which my wife and I keep canned goods. I see my hammer lying on the bench. I had forgotten to put it away on the tool rack. The hammer has nothing to do with why I came to the work room. I had come to get a can of mushrooms which my wife had asked me to bring to the kitchen for her. Seeing the hammer, however, brings to mind, besides the thought that I had not put the hammer away, a recent occasion when I had seen a friend using a similar hammer, what he and I had talked about as worked, the place where he was working, and the emotions he displayed when a misdirected blow of his hammer bent a nail. I am also briefly reminded of other things:

what I had last used the hammer for, for instance. The hammer I just saw shared several properties with items held in my memory. It belonged to the general class of hammers and to the broader category of "tools." It also belonged to the class "things my friend and I both use." The classes or categories related to my seeing the hammer include things seen or heard on a particular occasion and related emotions, as displayed by my friend when he bent the nail. I might also recall emotions I had felt in the past when I bent a nail.

This model suggests that the screening process scans multiple memory bins constantly as new input signals arrive and that some parts of screening activity – the identifying of input code units as similar to elements in memory, for example – are like, or perhaps the same as the links which join association elements. Directly after I saw my hammer, a code unit signal moved to my collection of memory bins. This signal activated some of the total number of bins. Bins activated included those containing images of hammers and their uses; the memories of the occasions on which I had watched my friend using his hammer; memories of the displays of emotion I had observed in my friend as he worked, and of how I had felt, observing my friend's annoyance when he bent the nail; and possibly several other bins. When the triggering signal reached each of these specially selected bins it acted as a stimulus to which the bins responded by sending a code unit signal representing their contents toward the attention field and on to other bins in which other related (associated) elements were stored. The secondary target bins then were stimulated to send signals to consciousness or on to tertiary target bins, and so forth. This sequential process clearly has its limits, of course. Our attention fields are not bombarded by a never-ending barrage of evoked memories. Some part of the screening process must set limits on signal travel, perhaps by itself responding to newer input, which new set of signals redirects screening activity. Further, the triggering input may itself be incorporated into a new association.

At some point along the chain of memory bins excited by an association triggering signal, the content bin's signals are directed, in addition to being directed to the attention field and secondary memory bins in the association chain or channel, to one of several feeling state hold-and-send storage sites. With some input signals (but perhaps not all), those signals go to content memory bins first, after which the content bins' responses provide signals to feeling state hold-and-send loci. The result of all this input scanning and stimulus-response activity is that one or several association events occur. I became conscious of the facts and even memories (content) as described and of the emotions I had had in the past, when I saw my hammer on my workbench.

For a triggering stimulus to produce the effects just described, it must be noticed, it must pass through initial input screening. Something determines what we pay attention to and what we later remember. We cannot take conscious note of everything we see, hear and touch.

A doctor and his wife are walking together along a sidewalk. A woman wearing a boldly patterned coat passes them, limping slightly. The doctor says, "did you see the way that woman was limping?" the wife answers, "No, but I did notice the really ugly coat she was wearing." Attention bias has been a factor in what each of these two people noticed. Input screening worked in one way for the husband and in a different way for his wife. Both of them saw the same thing, but what each of them noticed was different. The input each of them perceived had the potential to be an association event-triggering stimulus. The differences in what each of these two became aware were almost certainly the results of what each of them had in their individual memory bins as a result of past experiences, and of the effects the stored memories had had on their screening processes. It is notable that the perceptions of the man and his wife had feeling-state elements, necessarily due to the effects of earlier association formation: disapproval in the case of the wife and the memory of the satisfaction felt by husband in his having solved medical problems as a result of his having made careful clinical observations.

What input screening admits to the circuits activating association events is not the only thing which makes input noticeable and memorable. Some input is intrinsically noticeable and is naturally capable of producing association formation and retention. The hearing of an unexpected, suddenly occurring loud sound; the pain experienced with the touch of a painfully hot object; the sense of smothering which occurs when a person has unintentionally fallen into water deep enough to cover his face; and the feeling of falling itself are all instances of this effect. They all instantly occupy a person's attention, they are all likely to be remembered, at least for a time, and they all lead to association formation and usually to retention of the associations formed. Each of these attention getting types of input generate input signals which go to both content holding memory bins and to feeling state hold-and-send sites (to which target this input goes first is not always clear; the emotion resulting from a suddenly felt pain often seems to occur instantaneously following the pain producing event). Intense pleasant and welcome input, like the taste of some unusually flavorful food, also seems to capture our attention at once and to activate both content and emotion sites and circuits. The diner is likely to remember a pleasing meal if he sees the kind of food he had enjoyed previously listed on the menu of a restaurant. (As an aside, it is notable that these events and responses in our brains are dependent in large part on our sensory systems. A deaf man does not notice a thunderclap, unless he feels the vibration in the air around him produced by it.)

EXPECTATIONS

An important feature of the operation of input screening is that expectations are formed (and remembered). When I see a plate of food in front of me, and I recognize it as like a dish I have enjoyed in the past, I anticipate that eating it will bring me pleasure or satisfaction. This expectation is related to my associating the appearance of the food with its appealing taste, as a result of past experience. Expectations result largely from the ways the screening process classifies the things we see, hear, feel and think about. Screening labels input code units in several ways. We act as though the first step in classification (often, if not always) is to determine if input code units are familiar or unfamiliar. Input which is briefly experienced, not intense and not repeated may not loom large enough, or remain long enough in the attention field to displace what is already going on there. The student, referred to earlier, working in distracting surrounding may not take note of a briefly audible unfamiliar sound which occurs only once, although he might become aware of it in a quiet environment, if he was not concentrating on the material he was studying, or if it was repeated over and over. Novel input which is intense, which lasts long enough or recurs often enough, can capture a person's attention and provoke an exploratory impulse. "What was that clatter I heard just now? I wonder if..." Here, wondering is concerned with the prospects of future pleasure, pain, embarrassment or satisfaction. We act as though exploration after novel input allows us to classify that input and to develop a set of expectations.

When input is not classed as unfamiliar, this means that it relates to something in memory. Input code units may contribute to a usually unthinking activity like walking, they may be recognized as useful for completion of an ongoing project like finishing a training course, as helpful in performing a routine activity ("how much more do I have to do to get done with cleaning the room?"), or as useful in planning how to avoid some disagreeable experience. In all these examples, what is in memory is based on associations made up of both content elements, like visual images, odors and emotions; and feeling state elements.

LINKS: CATEGORIES

We now come to the task of identifying what determines the links – or neural channels – that connect the elements in associations. Some of the factors which govern the forming of association element links are part of what we all take for granted about the way we think. Other influences may be less generally obvious. What forms associations to begin with, what triggers association events, and how input screening and its memory bin scanning activity operates must all be closely related. A major member of the group of perceptions which link or connect association elements is the recognition that different elements share features which allow them to be seen as members of a common category. This link was referred to earlier in these notes in the discussions of input screening, of the initiation of association events and of the forming of expectations. There is general acknowledgement of the importance of this part of our cognitive activities. The section in a widely used dictionary concerned with the word "like" testifies to this fact. "Like father, like son" and "your necklace is like mine" appear in this entry in the dictionary as illustrating the meaning of this word, often made use of to indicate common class membership of two or more entities. There may be similarity between elements with respect to a variety of aspects: visual appearance (round, blue, large, etc.), where particular events occurred, and what feeling states the events produce, for example. Classes may also be defined by similarities between abstract concepts, like geometric shapes or ethical ideas ("moral," "sinful," etc.). We may formulate categories consciously in the course of performing mathematical or logical operations of course, but much of the category-defining process our brains perform is immediate, automatic and instantaneous. Further, we do not decide whether or not we will classify code unit signals. The classifying of input and other code units just happens. We cannot keep from categorizing things, and this property of our brains is probably necessary for our survival, which depends in large part on input screening ("that looks like a dangerous thing to do"), which screening must depend on our having formed and retained category definitions.

An additional feature of development in the forming of associations is that there are differences in the numbers and defining standards of the category repertoires of different individuals. The different responses of the doctor and his wife to the sight of the limping, tastelessly dressed woman illustrates one aspect of this set of differences. Another aspect is that people do not all have the same numbers of reference categories in their memory bins. A young child necessarily has less than an adult. A newborn infant initially has none at all, from a memory-cognitive standpoint. Someone who has lived his whole life in one small town may well have a smaller number of categories in store than a person who has traveled widely and who has learned one or more foreign languages, especially if he is illiterate or has no access to a television set, a radio or a motion picture theater. Commonly used tests of intelligence implicitly acknowledge differences in category numbers in storage in their use of questions such as "how is item 'A' like item 'B'" and in relating test responses to "mental age." Additionally, there are differences between individuals with respect to what code units come to be placed in what categories. A paranoid schizophrenic may take a simple "hello" from an acquaintance as a threat rather than as a friendly greeting, for example. Despite the fact that we vary widely with respect to the categories we have and hold, some basic categories – however we use them – must be essentially universal: familiar, unfamiliar; possibly dangerous; likely to be enjoyable, and so forth.

LINKS: TIME FACTORS

Time and timing factors are major influences in how associations are formed; stored and evoked (these associations are not entirely separable from those related to category development). There are several ways that the timing of code unit element arrival at the attention field, and at other functional domains to which they go or through which they pass, affects association formation. Input signals arriving simultaneously at a functional site must be at least momentarily associated. The different parts of a visual image are inevitably associated. The image, as it reaches the attention field, is a structured pattern, assembled out of the code units formed in single points of origin. In the case of sensory input, these points of generation are individual retinal sensors, touch responsive nerve endings in the skin, hair cells in the inner ears or somato-sensory or visceral nerves. What is known about the "unimodal" association areas in the cerebral cortex, and their location adjacent to the primary sensory cortical areas which receive signals from the visual, auditory and somato-sensory input systems, argues that interaction between the primary domains and the unimodal areas is responsible for the forming of the distinctive patterns, the existence of which make it possible to recognize the appearance of a friend's face or now a hammer feels when it is grasped. Beside formation of these unimodal input assemblies (associations), associations occur between the different types of input code units. A man walking across an uncarpeted floor sees the floor, hears the sound of his footsteps and feels the sensations coming from his feet with each step. In this experience there is at least a brief association between the input signals generated in three principal sensory domains. A second timing-related factor in development of associations is the order in which input signals are received. Classical conditioning procedures, pioneered by Ivan Pavlov, illustrate this effect. In Pavlov's experiments, a dog is presented with a distinctive stimulus such as a sound; then, after an interval, it is fed. Before the conditioning procedure began, it may or may not have heard that particular sound. If the dog had heard it earlier, its occurrence had not been followed by any notable pleasant or unpleasant experiences. However after the sound-feeding sequence has occurred several times, presenting the dog with the sound is followed by the dog's beginning to salivate. An association has been established, remembered and has become the basis for development of an expectation.

The time interval between the hearing of the sound and the feeding is important for the dog in determining whether or not an association is formed. When stimulus #2 follows stimulus #1 by too long an interval, no association develops. If the dog hears the sound in the morning and is fed at the end of the afternoon, many input elements come to the dog's attention between these two occurrences. In addition to this distraction effect, which results in there being multiple input signals arriving before the feeding, any one of which might become possibly associated with it, the long time interval between the sound and the meal can allow the sound to fade from short term memory before the food is given to the dog. When a person looks up a telephone number in a directory, as when he means to telephone a restaurant to make a reservation, he does not ordinarily remember the number for very long after he has made the call, unless his call is to a restaurant to which he has gone in the past, in which case the present use of the number has a different status than it would have if it was being looked up for the first time. In his making a second or third call to this restaurant, the person is effectively rehearsing a previous action rather than beginning a new one. (See later notes concerning repetition as an aid to association development and memory.)

Another time-related factor in association development is the epoch effect. Things experienced during a particular vacation or on a memorable holiday gathering are connected with each other in our memories (note here the overlap with category-forming processes). A person thinking about his senior year in high school is reminded of the activities he pursued during that year and of at least some of the people with whom he shared those activities.

LINKS: REPETITION

The role of repetition in the development of associations was pointed out earlier in these notes in connection with conditioning procedures. Several repetitions of the sequential presentation of a particular sound (the "conditioned stimulus") and the provision of food to the dogs (the "unconditioned stimulus" had to occur before the dogs learned to expect a feeding after hearing the sound. That this learning occurred makes it clear that the dogs acquired new memories during conditioning. Repetition and rehearsal are widely recognized as important features of learning new material and in maintaining what has been learned. Students learned the multiplication tables, before the availability of electronic calculators, largely by repetitive drilling. A violinist reads and plays a musical selection he is trying to memorize several times. Further, he practices the selection before playing it in a concert.

Repetitive drilling with the student trying to master arithmetic and with the musician is conscious and purposeful. Input repetition also acts in the forming and maintaining the memories of associations in situations in which it is spontaneous, coincidental and perhaps not fully conscious. This is well illustrated in the ways forgetting occurs. I no longer remember the telephone number of the home in which I lived thirty years ago. While I was living there I recalled it easily; there were many occasions to call it to mind when I needed to tell people how to call me, or when I telephoned my wife to tell her I expected to work later than I had expected. Since moving to my present residence (in a different city), there has been no need to think of that telephone number and I cannot remember it, although I might have a sense of partial recognition if I heard it mentioned by a member of my family.

Lack of repetition, besides allowing memories in general to fade, may extinguish an association. What is referred to as "extinction" in the writings concerned with classical conditioning involves presenting an experimental animal repeatedly with a stimulus (distinctive sound, light) previously associated with an unconditioned one (feeding, pain), without then delivering the unconditioned part of the conditioning pair. This is not the same as forgetting the association; it involves new learning. The animal remembers that the conditioned stimulus did not signal that food or pain were in prospect. The dog no longer expects to be fed when it hears the auditory signal which earlier had announced that a meal was on the way and it no longer drools when it hears the sound. This is a special instance of the effect or repetition: repeated presentation of a stimulus signals that something is probably not about to happen.

LINKS: FEELING STATES

A major group of influences affecting the forming and placement in memory of associations is that concerned with emotions. Some kind of feeling state is part of virtually all conscious experience, although much of the time, when the feeling state blend does not dominate our attention, we do not think about how we are feeling.

Many repetitions may not be needed to establish some associations. A man walking on a hillside path feels his foot skid sideways when he has stepped on a patch of loose gravel. The slope of the hill below the path is steep and the man sees that, if he should fall from the path he would be likely to be injured. He feels a brief burst of intense anxiety and responds to the situation by shifting his body forward and putting his other foot on a firmer part of the path's surface. He will not have to remind himself over and over that areas of loose gravel may call for his being particularly careful during the rest of his hike. He will include in the association just formed between loose gravel and risk of injury, sudden fear, and the sense of relief he felt after he succeeded in avoiding falling. The forming of the association cluster here was enhanced because strong emotion was one of its elements. This effect resembles sensitization effects in conditioning procedures.

The post-traumatic stress disorder (PTSD) is a further example of the effects of emotion on the development of associations. In this disorder, and probably in the case of the hiker, the forming of association pairs or of the element cluster and the entry of these associations into short-term memory is followed by naturally occurring reminders. These provide repetition effects for fixing the newly-formed associations in intermediate or long-term memory before they have faded from short-term memory. The reminders, besides reinforcing the memories of the associations and their elements are also triggers for association events. The hiker is likely to be on the lookout for other places on his path where there may be bad footing and is reminded of the recent near accident by the sight of areas which resemble the place where he nearly fell, and of the anxiety he felt at the same time. A person with PTSD, encountering a situation with – perhaps trivial – similarity to the one in which his disorder began, can re-experience the traumatic episode in memory, and the fear he felt because of the incident.

Some associations have properties which suggests that their existence may not depend on experience, but rather that they are essentially pre-formed because of the nature of the brain.

The startle response – which we think of as a sign of anxiety – resembles the Moro reflex. The infant jerks and moves his arms and legs abruptly when someone strikes the mattress of this crib forcefully, without warning. A person driving his car hears a loud crashing sound. He has a flash of anxiety and looks around. The anxiety is replaced by a sense of relief when he sees that the sound had been due to the striking of a wrecking ball against a building well away from the side of the road and not to some possibly dangerous occurrence on the road near him. A baby laughs when it father tickles it. He is not laughing in response to hearing a joke but because the tickling feeling has been associated with, and has provoked an emotional response (see also earlier mention in the discussion of association elements, of essentially inborn, pre-formed associations between emotion and pain or the sense of smothering).

LINKS: LAST THOUGHTS

The foregoing discussion argues that the influences which cause the associations to form, which trigger the occurrence of association events, which place and fix associations in memory and which generate expectations are intimately related. It may not be possible to disentangle these forces and influences completely. It is probably best to view them as different aspects of a single, integrated, data acquisition and processing phenomenon. Adding to the complexity of the consideration of association links is the evident fact that more than one type of link may connect elements in any given association cluster or in the course of any given association event. Data acquisition and processing activities must be going on constantly in all of our waking lives (and perhaps some of the time when we are asleep; we awaken when we need to empty our bladders, after all). Whether we intend to acquire and process data or not, we cannot avoid the flow of sensory input which impacts us. Our survival demands that we process input as it arrives. So association forming and the activation of association chains and association events must be occurring constantly in an automatic and spontaneous fashion.

SUMMARY: SOME FURTHER DISCUSSION AND CONCLUSIONS

What has been written here illustrates the difficulty of any attempt to understand associations. Formulating a simple and comprehensive assessment of the subject may be impossible. Any conclusions reached are sure to be provisional and tentative. We encounter new facts constantly, and we find new ways of evaluating those facts as experience grows. However, what has been discussed here has at least partly met for the writer the goals of the exploratory impulse which caused this analysis to be begun. Several aspects of the association process not often discussed have been exposed. That associations are experienced as discrete elements is one of these. Each association event, occurring as it does in an episodic fashion, has the quality of a response to a stimulus. Something has triggered the activation or the association, so that one or more of its components comes to consciousness, the attention, action planning field. Such triggers necessarily have some relationship or link to an association evoked or to one or several of its elements. Also, an association event trigger must be some kind of input signal a generated in the course of on-going experience: sensory material, or possibly ideas generated by a cognitive activity like the solving of a mathematical or logical problem. Some triggers may be input due to coincidental, unexpected occurrences. Some may result from deliberate, planned, intentional cognitive or overt activities of the person who is about to experience an event.

Associations are composed of distinctive elements and of the links between them. Individual elements exist in our brains as schemata, distinctive digital codes. These code units are retained in our memory bins and are capable of being transmitted from one functional domain to another – from memory to the attention field, from one memory bin to another, or from a memory bin to a site in which code units representing emotions are held and from which these feeling state code units may be sent to the attention field, to centers which control respiratory activity, heart rate, and other physiologic processes and to sites where urges to display emotions originate.

The channels through which code unit signals pass and the links which connect association elements, are specialized and selective. Any single input signal follows a track determined by what it represents and by the link most adapted to transmitting it. A painful sensation is more likely to be connected with an anxiety hold-and-send functional site than to a memory bin in which code units which stand for visual images of friends are stored. An implication of all this is that association elements can be thought of as falling into two broad classes. One class is made up of bits of information: visual images, word sounds, abstract ideas, various physical sensations, and so forth. The second class of elements differs from this first content-depicting class in that it is made up of the various emotion-related codes.

What association links or pathways are available for the development of associations, what tools are used to connect elements, are built-n features of our brains. They must relate to how associations form and how the activation of one association element brings to attention other elements. This group of processes must involve both signal travel from one site to one or more others and the responses of one or more code unit hold-and-send storage loci (memory bins and feeling state centers or systems) to the receipt of stimulant signals. Several association forming and retaining influences operate (we must keep in mind that associations themselves are memories). Among those are the input elements in spaced before-after sequences; receipt of the repeated signal of separate input elements consistently at the same time ("whenever I see John, Ann is with him")' input involving or capturing (by association) strong emotion; input which is very intense (very loud, very cold, very bright); and things perceived as appearing to share some special property or quality, which thus are seen to belong to a common class or category. That this last factor affects association formation reminds us that association building is how we learn to categorize objects, happenings and sensations in general. Of course more than one of these forces may operate in association formation and retention in any particular association-building episode.

Association building can only occur if the elements involved are held at least briefly in short-term memory before they fade, or go on to longer term memory storage or to some other functional domain, as both logic and daily experience demonstrate. The forming of patterns composed of input elements, including the assembly of association links and elements to produce completed associations cannot be instantaneous. It is a cognitive activity requiring synaptic activity and signal travel, both of which are known to require finite (and measurable) time intervals for their completion. Code units must stay in the attention field and in the functionally adjacent very short-term and moderately short term memory bins until the association process is completed. I have just purchased a flashlight and brought it home in a bag in which there are also two batteries I bought because the packaging of the flashlight was marked "batteries not included." I take the flashlight out of the bag and inspect it to see where the batteries must be inserted in it. At this point, the batteries are still in the bag, out of sight. Then I reach into the bag, remove the batteries from it and put them into the flashlight. The flashlight itself is in view during the whole series of actions; the batteries come out of the bag and into my field of vision as the task is brought to its end.

Additional properties of association assembly are that it is both automatic and continuously ongoing (and may be partly or almost completely unconscious). We do not deliberately decide to construct associations; the process just happens, whether we want it to or not. During study, research or analytic activities it is willed and purposeful, but much of the time it is driven by coincidence and the sensory input of the moment. What elements come to be components in particular associations depends in large part on input screening, which in turn determines what sensory information reaches the attention field (and other domains where input leads to and structures responses) and what comes to attention from memory bins and emotion-related element storage sites. Screening itself necessarily makes use of the memories or previously formed associations developed over an individual's life span.

Input screening has several additional features (or consequences). Input code units go to several destinations. Input may go to very short-term memory for use in organizing and completing an action in process, like the pronouncing of a word (such input soon fades away). It may be held in intermediate-term memory for use in performing activities of more than momentary duration (minutes to years), like adding a column of figures or working to earn a college degree. It may also go to very long-term memory, which differs from the other memory sub-domains in that it appears to involve or depend on micro-anatomic changes in neurons (and perhaps macro-anatomic changes as well, (as evidenced by the differences observed in the anatomy of the occipital cortex in persons blind from birth as compared with individuals with normal vision). Very long-term memories generally and memories of and for associations generated in infancy and early childhood seem to be lost only with destruction of the storage sites (bins?) by cerebral infarction or processes like Alzheimer's disease.

END NOTES

It is significant that a code unit element can occupy more than one memory or cognitive subdomain at a time. I can think about the fact that two plus two equals four, can use that fact in an ongoing calculation and at the same time hold that fact in long-term memory.

Some further considerations came to mind during the writing of this study. The first is that, given the natures of neurons and of the brain in general and the ways signals move to and within the brain, it is not likely that any association element can be without links or connections with one or more other elements. All association elements are latent or in storage in some sense between the occurrences of the association events which include them. This notion applies to both experience-derived elements in memory bins and to code units held in the several feeling state-related loci. Sensory systems send signals of different kinds (visual images, sounds, touch sensations, et cetera) constantly towards consciousness where the signals appear simultaneously or in distinctive and closely spaced sequences (which sequences may have, each one, a special organization). Some associations are experienced as though they were inborn like the feeling state responsible for an infant's laughing when it experiences tickling sensations.

A further consideration concerns memory. It is one generally, and properly, taken for granted. It is discussed here because of its relevance to input screening, association event triggering, association-formation and association retention in memory. Any input signal moving toward the attention field, coming either from primary sensory sources or from in-brain loci like the memory bins has a finite and characteristic duration of residence in consciousness and in the short term memory sub-domains serving the planning of actions in process. This duration effect also applies to persistence of code units in sub-domains which organize and regulate actions we do not ordinarily plan and carry out consciously and deliberately, like picking up a coffee cup or rising from a seated to a standing position. This persistence does not depend on rehearsal or other memory-enhancing influences. I recall easily what I had for dinner last night but I do not recall what I ate for my evening meal a week ago. It is as if input code units have a natural tendency to be held on short-term memory for a time (which time interval may relate to the nature of the code units and to the circumstances surrounding their generation), after which they fade away with respect to the attention field or other input response processing domains to which they have come. This notion is supported both by logic and by everyday experience. If we did not discard much, perhaps most, of input code unit signaling, our attention fields would become overloaded and chaotic, and the data processing functions of our brains would become unable to cope because of the need to determine which, out of an enormous number of elements held in the various memory bins, should be used in responding to what is going on around us.

ADDENDUM: REGARDING NUMBERS OF ASSOCIATIONS HELD IN MEMORY

The total number of associations retained in any person's memory is determined by the number of code units in storage. In a very simple brain, with only two elements retained, there could be only a single association. A brain holding three code units might have three two-way associations and one triadic association for a total of four. In theory, the number of associations could increase exponentially, with increasing numbers of stored elements. The number of feeling-state-related elements is necessarily mostly fixed, since it depends on the intrinsic properties of the brain. Elements in memory bins, on the other hand, grow in number with the accumulation of life experiences. A child entering the first grade must have fewer association elements in his memory bins than he will have at high school graduation.

The number of associations actually formed from the total store of code unit elements is less than the theoretical maximum. The nature and function of association links ensures that there is selectivity in association assembly. Every element is not associated with every other one. When I see a fried egg on my plate at breakfast, that image does not bring to my mind's eye the face of my father. The situation is further complicated by the fact that items in memory may be lost through non-rehearsal or with brain damage.

SECOND ADDENDUM: REGARDING THE MAINTAINING OF ASSOCIATION ELEMENTS AND ASSOCIATIONS IN THE ATTENTION FIELD DURING ASSOCIATION EVENTS AND OTHER COGNITIVE ACTIVITY; ALSO, SOME NOTES CONCERNING MEMORY

Holding code units in the attention field during association events and other cognitive activity cannot be analogous to hanging a picture on a wall. The mental pictures of what occupies consciousness are not fixed and static. What goes on in the brain when a person is thinking about something must be due to some type of neuronal activity. Such (any) neuronal activity consists of the activation of neurons, with transmission of signals from cell to cell or along chains of neurons (see earlier portions of these ruminations). Keeping an item in view during an association event or other cognitive activity must make use of a process which in many ways resembles that which makes the image of a person on a motion picture screen seem to be present continuously. What a viewer sees seems to be constantly present, while in fact it is a series of images, projected so rapidly that his eyes and brain do not recognize the individual elements of the series (unless the rate of projection of the images slows down, allowing the discontinuity of the members of the series to become apparent). With television or moving pictures recorded on discs, what is sent to a screen, instead of being a series of projected photographs, is a series of coded signals, spaced as with the motion picture so the object represented appears to be continuously present. We do not have screens like those of the older motion pictures, of television sets or of computers physically embedded in our brains, but it is clear that there is something in the brain which has a function analogous to that of such screens. This notion implies that the holding of an image or other code unit in consciousness involves a repetitive sending of signals to the attention field as we visualize or think about something. Some of the sources of signals repeatedly transmitted to the attention field when stimulated necessarily are memory bins (see later notes concerning persistence of feeling state perceptions in the attention field – Third Addendum).

Several lines of evidence support this idea. "Mirror neurons" are found to be active in the frontal and parietal lobes of a monkey when it sees another monkey performing some action associated with increased activity in those same regions of the brain (thought to be involved in planning and regulation of the action), even though the watcher is not itself carrying out any observable movements. It is as though the animal which sees the action is thinking about it (see references to electroencephalographic data in Kandel, et al.). Such signs of increased cerebral cortical activity point to the presence of neuron-to-neuron signal transfer, considering what neurons do. The facts that such signaling includes synaptic events and that the signs of increased neuronal activity are present over a long enough time to detectable by electroencephalographic and other data recording systems (see paper by Schulte-Ruther et al in the Journal of Cognitive Neuro-Science, listed with the sources for this study), it is virtually certain that repetitive signaling is taking place during periods when mirror neurons' activity is increased (increased above baseline activity).

The mechanisms which keep code units in the attention field and in short-term memory differ from those holding information in long-term memory. Long-term memory does not appear to depend on expenditure of energy above that of neurons' resting baseline (for example, that of slow wave sleep). Several observations support this idea. To begin with, the mirror neuron experiments just mentioned demonstrate that an animal observing another's performing an action has an increase in localized cerebral blood flow (and thus, by inference, an increase in metabolic activity related to increased cellular activity) in brain areas known to be involved in programming and regulating the observed motor activity. This indicates that, before the watcher saw the animal it was observing start to move, its programming neurons were less active than they came to be after the observed animals' actions began. Further, the maintaining of information in intermediate and long-term memory does not appear to be critically energy dependent. Individuals who come out of deep barbiturate-induced come, during which cerebral metabolic activity is profoundly reduced (the EEG may record no electrical activity at all), do not typically have loss of memory for the meanings of words or for the major past events of their lives. It is plausible to think of the processes which keep association elements and the associations of which they are components in the attention field are different from those involved in long-term memory storage. In their lack of need for ongoing energy expenditure to maintain items in long-term storage, the brain's memory bins resemble printed pages, which have been changed by the printing processes. The pages are no longer blank, as they were to begin with, and what is printed will remain on them unaltered unless it is erased by an active process, or the pages are destroyed. There are reasons to believe that the brain's memory bins are physically altered when long-term memories are established (see Kandel, et al), and so have properties like those of pages. Further, they give back the information they hold (a memory comes to mind) when stimulated, as pages do when someone reads them.

THIRD ADDENDUM: CONCERNING FEELING STATE SEND AND/ OR HOLD LOCI

It was pointed out earlier that these loci (for example the well-defined reward locus or system/network) act as though they become active and send signals to the attention field (AF) when stimulated. Such stimulation must be internal; a result of primary sensory input signal travel or as the consequence of the arrival of signals coming to the feeling state locus or loci through association chains or pathways. The point of this addendum is to analyze the fact that feeling states, as perceived in the AF, have duration; they are felt over varying time periods from seconds to days, weeks, or longer (perhaps the longest durations are due to repetition or rehearsal effects, operating along the association chains which end in or send stimuli to the feeling state loci).

In an earlier addendum it was proposed that one way an element can persist in the attention field is its repeated delivery at short intervals, analogous to the way an image appears to persist on a television screen when repeated code sequences are sent in an organized pattern (and analogous to the apparent persistence of an image on a moving picture screen when images are repeatedly projected onto it).

If a repeated projection mechanism is responsible for the persistence of feeling states in the AF, there are several ways in which this might occur. One mechanism would be the continuing sending of a feeling state signal (reward, fear, et cetera) from the feeling state hold-and-send locus or system once the locus has received a stimulus, an activating signal. The response of the locus might be vigorous at first and then gradually become less vigorous, and finally fade away, as the movement of a pendulum eventually stops if the displacement of its position which set it in motion is not repeated. A strong stimulus would lead to a longer persistence of the locus' transmission than a weaker one. This kind of effect is plausible; powerful fear inducing experiences have effects which are more enduring than minor ones. The flash of anxiety felt with a misstep while walking on an uneven surface, when there is a momentary fear of falling, is quickly gone with the recovery of secure balance. A walker who has nearly been hit by a passing car, on the other hand, may feel shakiness and sweating for several minutes or longer. Intensity of the emotion signal transmissions we experience clearly does vary, as does their duration; the feeling states we experience do appear with strong or moderate intensity and then gradually fade. The feeling of well-being which starts during a satisfying meal, sexual intercourse or as a result of an agreeable conversation with a friend continues well beyond what induced it, before gradually fading away.

The just described hypothetical mechanics must involve memory, both as suggested in the notion of a temporarily activated para-attention field site and in the primary emotion-related sites. Repetition or rehearsal of whatever activated the feeling state hold-and-send system in the first place makes the feeling state appear again in the AF (witness the PTSD). So feelings stay longer in working memory ready to be sent to the AF if what induced them is repeated or continued. Further rehearsals or reminders of past emotion-inducing experiences, seem to bring the emotions easily to mind (again, think of PTSD). Lack of rehearsal leads to forgetting and loss of the feeling state perception, except possibly with associations involving feeling states established in infancy or early childhood.

LAST ADDENDUM: CONCERNING DISPLAY OF EMOTIONS

An association event is a response, or a series of responses, to the input which triggered it. Preceding each overt, observable or cognitive, covert, response there must be an urge or tendency to take action, without which the programming and execution of the responses cannot take place. This urge or impulse to act must appear because of how the brain's circuitry is organized and on the connecting of input via input screening to memories of past experiences, including their emotional components and to expectations ("If I drop this glass on the floor, it will probably break"). Such signal travel in the brain necessarily depends on activation of association chains.

The simplest example of the relations between input, input screening, and response action is the deep tendon reflex. The knee-jerk movement occurs because of the ways the sensory and motor neurons involved are made, how they are placed in the spinal cord, and with how they interact. The urge or impulse of the motor nerve to act, to "fire" is inherent in its nature. If it receives a certain kind of stimulus, it must respond. Input screening involves the types of sensory nerves stimulated and the response threshold of stretch receptors in the patellar tendon. A light touch delivered to the skin over the tendon produces no response, nor does a weak tapping of the same area. Such input is screened out; a stronger blow, sufficient to affect the stretch receptors of the tendon, produces a knee jerk.

Actions more complex than tendon reflexes, and the urges or forces causing them to begin, must also be induced to occur by input signals which survive input screening, but they are still structured by the intrinsic properties of the nervous system. Examples of such responses are the stepping movements of new-born infants or of "spinal" (brain-stem transected) cats. In these instances more signal travel through neural circuitry occurs than in simple deep-tendon reflexes, but input screening is still simple and automatic, probably based on such factors as body position (upright versus horizontal, or vice versa) or foot contact with a surface.

With still more complex behaviors like the display of emotion or the beginning of exploratory activity after receipt of unfamiliar input, responses still begin with input, followed by input screening. Then there is, with signals which survive screening, development of an urge or tendency to act, and this leads in turn to response programming and finally to muscular (or cognitive) action. The development of an urge or urge-equivalent impulse to act, a choice of which of a number of action programs in storage is to be employed and actual performance of the response are all components in the response process. Action programs can be inborn, like reflexes, or developed through experience, like skills.

Impulses to action we all share include the urge to escape from pain, the tendency to act to prolong pleasurable experiences, and the impulse to express emotion. These all must all depend on signals traveling along one or several association pathways which – for most, if not all actions – involve activation of feeling state hold-and-send functional domains. A way of conceptualizing the sequence of events leading to a display of emotion is this: first, there is primary input; then input screening occurs; then code units which survive input screening proceed to an "action to be considered" functional domain which is related to, or integrated with an urge or tendency to action; then signals must go to a response planning domain, in which a choice from the list of possible action programs is made. After all this, signals go to loci (probably mostly prefrontal lobe sites) where specific actions are planned and from and from which stimuli are transmitted to the primary mother cortex of the frontal lobe (and perhaps to other brain sites where "thinking" occurs); observable (or reportable, as with reasoning) actions then occur.

The repertoire of action programs involved in emotional display is relatively small and relatively specific to the feeling states portrayed: we smile with pleasure, laugh at a joke, jump with joy, or weep with sadness. Most, if not all, of these action programs are built into us, although the initial response urges and feeling state displays may give rise to complex behaviors which are extensions of emotional expression actions (below).

The action-generating targets of the feeling state hold-and-send domains, where action programs are stored when not in use, share with the neuro-endocrine response systems controlling pulse rates, sweating, stress hormone secretion, and so forth, that they produce feedback to the attention field. I know that I am smiling because of how my facial muscles feel. Action-generating target sites also resemble neuro-endocrine response systems in that the actions they organize and set in motion ordinarily occur in an automatic and involuntary way. A person being given a massage for relief of muscle soreness may smile and sigh with pleasure as his pulse and respiratory rates slow because of the relief of distress and the relaxation he is experiencing.

Behavioral and neuro-endocrine responses to emotion differ in that the urges to behavior and the actual behavior potentially following those urges can often be consciously inhibited or modified. A person following the advice to "turn the other cheek" after being slapped has blocked his (almost certainly present) urge to return the blow.

Beyond the immediate, spontaneous expression of emotion, like the smile at the unexpected sight of a friend, emotion is expressed in other ways as well. The smile was involuntary; the later handshake was a conscious, deliberate act. The actions indicating emotion in particular situations are clearly context-driven. What is happening determines how we feel and how we show our feelings. That this is true makes it clear that the ways we express our emotions are the results of the actions and interactions of complex chains of associations.

SOURCES

Corsini, R, et al. The Dictionary of Psychology Ballanger-Routledge (The Taylor and Francis Group), New York and London, 2002

Diagnostic and Statistical Manual of Mental Disorders, Fourth Edition. American Psychiatric Association (Copyright 1994), Washington, D.C.

Kandel, E; Schwartz, J.H., Jessel, T.M. Principles of Neural Science, Fourth Edition. McGraw-Hill, Health Professions Division. New York, St. Louis, San Francisco (and 18 more cities) 2000

Kandel E.R. In Search of Memory W.W. Norton and Co., New York and London, 2006

Kasper, D.L., Fauci, A.S., Longo, D.L., Braunwald, E., Hauser, S.L., Jameson, J.L. (editors). Harrison's Principles of Internal Medicine, sixteenth edition. McGraw-Hill Medical Publishing division, New York, Chicago, San Francisco, and 11 other cities, 2005

Pardee, N. E., and any reader of these notes, <u>The Widely Experienced Events of Everyday Life</u>, in production, publish date TBA

Piaget, J. <u>The Origins of Intelligence in Children</u>. N.W. Norton and Company, Inc., New York (published by arrangement with International Universities Press, Inc., Copyright holder after 1952)

Piaget, J., <u>Play, Dreams and Imitation in Childhood</u>, W.W. Norton and company, New York, London, 1962

Plum, F., Posner, J.B. <u>The Diagnosis of Stupor and Coma</u>, Third Edition. F.A. Davis Company, Philadelphia, 1980 and 1982

<u>Random House Webster's College Dictionary</u>, Random House, Inc., New York, 1991

Schulte-Ruther, M., Markowitsch, H. J., Fink, G. R., Piefke, M., <u>Mirror Neuron and Theory of Mind Mechanisms Involved in Face-to-Face Interactions: A Functional Magnetic Resonance Imaging Approach to Empathy</u>. Journal of Cognitive Neuroscience, 19:8, pp. 1354-1372, 2007

Darwin, C., <u>The Expression of the Emotions in Man and Animals</u>, Third edition. Oxford University Press, Oxford, New York, First published 1872, copyright (current edition), 1998

Caplan, F., <u>The First Twelve Months of Life, Your Baby's Growth by the Month</u>, A Bantam non-fiction book in association with Grosset and Dunlap, Inc., New York, Toronto, London, Sydney, Aukland. Originally published as twelve individual booklets, 1971; hardcover edition, 1973

"SHOULD," WITH NOTES ON GOALS, EXPECTATIONS, RULES, STANDARDS AND PRINCIPLES

DEFINITION AND GENERAL NATURE OF "SHOULD"

The Random House Webster's College Dictionary identifies "should" as an auxiliary verb. "Should" is defined as follows: (1) "Prefer it on past tense of shall'" (2) A word, "used to indicate duty, propriety or expediency (expediency in this context means, "advantageousness, suitability for a purpose, proper, conducive to an advantage, governed by self-interest, a handy means to an end" [from the definition of expediency from the same source]); (3) "Used to define a possible condition" (one imagined or considered but not necessarily presently existing); and (4) "Used to make a statement less direct or blunt."

"Should," as an auxiliary verb cannot stand alone. Its use must be associated with another verb. "To should" is entirely different from "to walk." "He should" is essentially meaningless, does not point to a specific action in the way "he walks" does. And "should" does not function like an adverb. "He walks shouldly" is a nonsense sentence. "He should walk" can make sense, but not in the way "he walks unsteadily" does. In the expression "he should walk," "should" brings the notion of desirability into the sentence, rather than a description of how the walking might be performed.

"He should walk" carries with it some additional implications. It suggests that a choice is open to the man being discussed; he might decide not to walk. This sentence also has an element of ambiguity. It might be made by an observer watching the man walk, in approval of the man's being active. It might also be made by someone considering the possibilities that the man might, at some future time, decide to walk or not to walk, when the person making the statement is considering what the man could be doing later, rather than what the potential walker is doing at the moment.

"Should" can also be used in relation to a verb-adverb complex. "He should walk on the sidewalk" has implications about where the man's walking should be done, rather than whether or not the man should walk in the first place. This statement also suggests that the man may have choices to make: in this case where he should walk (possibly on the sidewalk, on the street or away from the street, on the grassy strip beside the sidewalk) rather than whether he should walk or sit down. A rule or principle enters the picture here. "He should walk" carries with an undefined implication that, if the man walks, some benefit, to someone, will be gained." "He should walk on the sidewalk" must have its basis in a rule or principle, like "sidewalks are made for pedestrians so they don't have to walk in the street or on the lawn beside the sidewalk." The various settings in which "should" is used and the verb-adverb complexes with which it is associated in any particular use, relate to a number of factors: desirability of a thing or action, choices and alternatives, regard for rules or principles and imaginings about things or actions which could exist or occur at some future time, but which are not presently in view.

RULES, STANDARDS, PRINCIPLES, EXPECTATIONS AND GOALS

The recognition that "should" carries with it the implication that an action or set of conditions must conform to some standard, rule or principle if a benefit is to be expected leads to a need to consider the relationships of goals and expectations to rules and principles. Logic dictates that no rule, standard or procedure is formulated unless the rule-maker (or makers) has in mind some goal. Laws prohibiting theft aim to make people secure in retaining what they own. The rules of football make playing the game possible, so there can be easily identified winning and losing teams and so that no player is subjected to excessive risk of injury.

The goals leading to the creation of rules and standards are of two general types: (1) those which direct individuals to do or to value specific things (to work for wages, to refrain from stealing); and (2) those which are experienced or anticipated with respect to how a person, or the members of a group, will feel when specific goals are reached. These last goals include feelings of satisfaction (elation when a team wins a game), or when a project (perhaps the building of a house or planning of a garden) is successfully completed. These two kinds of goals might be thought of as first and second-order goals. The successful accomplishment of first-order goals is a necessary precursor to the reaching of second-order, emotion or feeling-state related ones.

First-order goals involve specific objects or experiences: a sprinter or competitive swimmer wins a race: a craftsman completes making a chair or footstool; a traveler sees a cathedral he has read about and has wanted to see. First-order goals are essentially infinite in number, considering the diversity of people's interests and desires. Second-order goals, the experiencing of wanted feeling states, are much fewer, and are more or less common to all individuals: satisfaction, comfort, the relief of pain or anxiety, the perception of being loved or approved of and so forth.

First and second-order goals differ in their relationships to rules, standards and laws. Reaching a first-order goal generally involves actions or experiences (seeing a cathedral a traveler has read about with interest requires that he look for it in the right city) guided or determined by a rule or principle. The pedestrian walks on a sidewalk instead of on the edge of a street or on the grass beside the sidewalk partly so he can avoid the risks inherent in walking on the street, and partly to avoid violating the rule stated on the sign which reads, "Don't walk on the grass." The second-order goal he reaches is the avoiding of anxiety. Avoiding of or the relief of anxiety are goals we all share. Such goals are part of our genetic heritage. They are latent, always ready to be invoked. Feelings of anxiety come to consciousness when what we observe or experience sends signals (through a chain of associations, presumably) to anxiety-generating loci or systems in the brain, which, while ordinarily more or less latent when not stimulated, then send other signals to consciousness. The elements in association chains which act to stimulate emotion-transmitting centers or systems are necessarily memories in storage, which in their turn, remain latent until the chain is activated by some input, some experience (a sound, sight, or physical sensation or a complex of input signals). In this kind of process, a (or the) goal, the relief of anxiety, is reached when anxiety-depicting signals diminish so that they do not reach consciousness, do not occupy a person's field of attention, and when "feel-good" signals appear in awareness, presumably because they are being sent by reward-related brain systems as a result of their being disinhibited after anxiety has diminished, or because the systems sending them have been stimulated by messages reaching them from one or several other association chain elements.

The pedestrian's reaching his second-order goal is due to his having chosen to walk on the sidewalk, but this results from his having reached the first-order goal, and this, in turn, has been due to what he has learned or observed.

Some first order goals are simply inappropriate because of the nature of things. The way things in our universe are put together generates some principles (notably, scientific principles which may be distinguished from rules made by humans or groups of humans, like laws enacted by legislatures). A woman sees a pair of shoes displayed on a shelf inside a store window, and finds their appearance appealing. After entering the store and talking to the clerk, she learns that the shoes she had seen were the only ones in stock there, and that they were size five. She knows she needs size seven shoes and she abandons the (tentative) goal of buying the pair on display, recognizing that, if she does buy them, she will not be able to wear them comfortably.

An example of how second-order goals might relate to first-order ones is what happens when a competitive swimmer wins a race. The envisioning of a goal has been followed by an effort to reach it, by use of actions and procedures prescribed by his coach (rules, etc.), and by succeeding. Second-order goals are met through several pathways or association chains: the "tried-to-do-it-and-succeeded" general sequence, leading to feelings of satisfaction; the approval of others – coach and teammates – leading to the pleasure of having been thought well of by others; and the relief of anxiety ("I'll feel awful if I lose this race").

Some second-order goals may be reached coincidentally, rather than by conscious effort. A driver may experience pleasure on seeing a picturesque vista, after his car comes around a bend in the road. Encountering the view had not been on the driver's mind when he started his trip, which had the goal of seeing an old friend living in the town toward which he was driving.

MORE ON RULES, STANDARDS AND PROCEDURES: ALSO, SOME
NOTES CONCERNING EXPECTATIONS

The types of standards associated with efforts to reach first-order goals, and thus with the use of "should" in this context, are many. They may be laws made by legislative bodies or edicts issued by political leaders. The may be religious principles or cultural traditions (what to believe, what is moral – what people do or refrain from doing, as members of a sect or social group). Some rules and principles evolve within families or neighborhoods. The interactions between infants and their parents lead to the infants' expectations and to what amounts to rules of behavior they must follow. A mother's smile can cause a one-year-old to expect a hug or a treat. A toddler of two has learned that when he awakens in the morning, he will soon be getting dressed. These learned patterns amount to rules about how a day "should" be organized. They are established in an infant's mind early enough so that the experiences which lead to their formation, and their "should," are not remembered as individual events (as attending an algebra class is recalled by a high school student); we learn a lot before we are three. Early learning of this kind resembles classical conditioning. Some moral and ethical rules are learned consciously, and the learning experience is remembered (often, though not always - see Jean Piaget's discussion of children learning the rules of the game of marbles). The need for players to take turns in performing the actions carried out in playing board games (checkers, etc.) and that students must stand in line, not pushing forward to more advanced positions in line to get ahead of other students when a teacher is handing out teaching materials to be used in an upcoming class fall into this category.

In addition to the experiences already mentioned, there is an important – and unavoidable – source of "shoulds:" direct daily experience. "The last time I ate something that looked like that it tasted great!" or, "When I tried to pick up one of those yesterday, I couldn't do it; it was too heavy." Thus expectations born of experience give rise to "shoulds."

It is theoretically possible that some rules and standards are inborn. If this were the case, it would be an exception, or group of exceptions to the idea that experience, and the expectations derived from it, give birth to most (if not all) explicit rules, standards and principles. An argument against the notion that we are born with preformed rules and standards embedded in our minds is that it is nearly impossible to imagine a use-event of "should" which would be completely unaffected by previous experience or by the circumstances of the occasion about which, or during which, "should" was said or written. Whatever beliefs or urges we may be born with, they can hardly be expressed in a particular situation without their being affected by what we have learned. An inclination to fairness, which might cause a child to share a toy or part of a candy bar with a playmate, might be present at birth, but knowledge about toys and their uses or candy and its appeal can enter his mind only as a result of experience.

Rules, standards, principles and expectations must be in storage somewhere when they are not being made use of by a person who says or writes "should." We assume, certainly correctly, that the storage sites are in human memory or in artificial data repositories like the pages of books or journals, computer memories or compact discs. An aspect of this storage not often discussed is that the storage systems hold the retained elements outside of consciousness when they are not actually in use (as in the case of most items which make up our memory stores, for that matter). Individual rules and standards are "out of mind" most of the time. Also, the elements are necessarily components in, or are closely linked to, chains of associations.

One problem associated with "should" calls for special attention. Different "shoulds" may be in conflict with each other. We generally agree that the revenues of the federal government "should" be in balance with government spending, and also that individuals who are ill or malnourished "should" have their needs met. Reports in the media, however, suggest that, at least at present, both "shoulds" cannot be satisfied.

ADDITIONAL NOTES ON EXPECTATIONS AND GOALS

Expectations and goals are closely linked. The development of a goal must depend on one or several expectations derived from experience. The emergence of a goal in a person's mind leads naturally to a consideration of how that goal might be reached, to what actions reaching it might require and thus to a review of what rules or principles might determine how those actions should be organized. The constitution of the United States can be thought of as a collection of rules, standards and procedures designed to make it possible for the goals stated in its preamble to be achieved. Some expectations are produced by classical conditioning. Certain behavioral rules are learned by kindergarten students so the goal of avoiding criticism by teachers and fellow students can be met.

Goals may be partly or mostly unconscious. A man does not ordinarily think about what he is doing when he steps around a patch of mud on a path, although, on later reflection, he may recognize the goals served by this action: not slipping on the mud and not having to spend time cleaning his shoes. It is worth noting here that the goals (for that particular instance of what he should do) form instantly and instinctively in response to the conditions of the moment. In the background of this act are other goals (ones present over long periods of time, like having clean shoes and not falling down), derived from experience.

REWARDS AND PENALTIES

The notion of rewards and penalties is basic to a further consideration of second-order goals. Second-order goals do not have to be formed through experience; they are present at birth and stay with us, at least in a latent sense, for our whole lives. We value feelings of happiness or satisfaction; we seek to avoid pain and anxiety. We enjoy pleasantly flavored food; here physical sensory input and the feeling state merge, although the sensory input occurs before the appearance of the feeling state; and this indicates the existence of an association chain. We are motivated to try to avoid or to eliminate physical pain, nausea, dizziness, smothering sensations and exposure to overwhelmingly intense light or noise, by avoidance or withdrawal.

The paths by which second-order goals are reached can be intrinsic and inborn (as with the pleasure associated with eating flavorful food) or they may be formed by learning (or both; they may, probably often do, involve both experience – derived and genetically determined factors). A newborn infant, experiencing hunger for the first time reacts automatically to the touch of a nipple on its face or lips, and sucking provides both sensory pleasure and (we infer, when watching the infant) satisfaction as hunger is relieved. Some (many?) inborn goals are less obvious and more complex. We, and rats in laboratories, appear to have an innate urge to explore unfamiliar things and environments. A rat placed in a new cage moves about in what appears to be exploratory activity, interrupting its usual grooming and feeding for a time, and sleeping less than was usual for it in its familiar cage. Eventually it returns to its previous activity pattern and to its earlier sleep-wake habits, as though a goal has been reached.

The rewards for reaching second-order goals might be thought of as coming from functional domains in the brain which, when stimulated, send welcome feeling state signals to consciousness. There must be a number of such domains. Feelings of happiness and of fear are distinctive (though different from physical sensations like warmth, pain, usual images or touch): we recognize and name them, and we do not experience all of them all of the time, even though we may be aware of mixtures of one or more feeling states, as when we feel both happy and excited.

The swimmer mentioned earlier wins a race. He had done the things his coach told him he should to prepare himself for it. The rewarding, wanted feeling states come to his attention field from somewhere in his brain. They are primary sensations like pain or touch. A train of stimulus-response events occurs so that one or several functional domains in the swimmer's nervous system is stimulated and then sends signals onward along an association chain or pathway, eventually to the domain which is so structured that, when it is stimulated, it sends feel-good signals to consciousness. Such chains or pathways must have several properties. They are distinctive and selective; each chain becomes active only when one of its elements (the first, presumably) receives a stimulus of a particular kind; the chains are activated and transmit signals automatically; they are a memory complex; they may also have both inborn and learned components.

That the processes like those just described influence how "should" develops is supported by laboratory observations involving rats and motivation. A rat which has had the tip of an electrode placed in a critical site in its fore-brain, and which finds that, by pressing a lever, it can activate the electrode, acts as though it experiences pleasure with firing of the electrode. It abandons its usual grooming and feeding and presses the lever over and over. The special site for electrode placement in such procedures is not along any of the recognized sensory input pathways (vision, hearing, touch, etc.). It acts as though it was part of – or at the end of – an association chain, the elements of which received stimuli from other elements and which responded to stimulation by sending signals on along the chain. The special electrode placement site responds to stimuli which reach it by sending signals to what must be the rat's version of consciousness, in which domain input leads to the planning and the performance of observable actions. Having a stimulus provided by the electrode seems to have the effect of the rat's reaching a second-order goal. This model includes several factors: experience leading to expectations: experience and expectations leading to discovery of procedures which, when followed, bring about successful accomplishment of first-order goals (pressing on the lever produces downward movement of it); and the satisfaction resulting from the firing of the electrode. So the rat learns it "should" press the lever if it wishes to see the lever move downward and to experience pleasure (expected, as a result of movement of the lever).

SUMMARY AND CONCLUSIONS

"Should" is referred to in the Random House Webster's College Dictionary as an auxiliary verb. It does not stand alone in any expression. "To should" is essentially meaningless in contrast to what is meant when a person says "to walk." It modifies the verbs with which it is associated by indicating the desirability or undesirability of an action, object or set of circumstances. Further, it does not function as an adverb, does not refer to how an action is performed or to the properties of an object. "Should" may occur in expressions concerning things or actions which are imagined, which may never have occurred or which may not be in prospect, though it also can be used in connection with actions currently in progress.

We commonly think, when we imagine ourselves involved in an activity or consider a particular object that the object or activity will have distinctive qualities which we value. Many, possibly most, such ideas arise out of experience; they relate to expectations. If we see "A" we expect that "B" will soon appear, since this has been true in the past. Some expectations result from easily recalled experiences; others, still born of experience, have their origins far enough in the past, or in early childhood, so that the events which led to their development have been lost to conscious memory. This is probably particularly true with respect to expectations generated in infancy or in the toddler years. A baby cries, and is then fed or cuddled. The person who was that baby does not remember the first time, or the first of many times, when that occurred. It is clear that many important things are learned early in life, before we have the mental tools to form discrete event memories.

Some expectations, which may in a sense be inborn, are those like the anticipation that the investigation of something unfamiliar – an object, an idea, a neighborhood – will lead to desirable effects, to the meeting of first-order, second-order, or both kinds of goals (it is likely that the achieving of a first-order goal will essentially always give rise to the meeting of one or more second-order ones). It is clear that the idea of goals is closely linked to that of expectations, but goals and expectations differ in important ways. A goal is, by definition, something wanted. An expectation, on the other hand, can be positive, negative or neutral ("I'd be really happy if I could do that;" or, "I'd hate it if that happened;" or "I don't really care about how that turns out").

Considering goals leads naturally to an attempt to determine whether or not they are reachable and to a review, if they are, of what procedures, rules, standards and principles would need to be taken into account in any effort to reach them. A man planning to paint the walls of a room in his home will try to visualize how the room will look when the job is finished. Both his personal preferences and those of his family members enter into the planning of the project. Some of these standards may be inherent or instinctive. We speak of bright, cheerful colors and contrast them with dark, somber ones. Other standards may be learned. The man's wife has told him that she likes rooms with pale yellow walls (he wants to please her. If she tells him she likes the appearance of the room after the painting is done, her expressions of approval will make him feel good).

"Should" also enters into the ways the painting of the room is planned and carried out. A hammer cannot be used to apply paint. Some kinds of paint are best for coating wooden surfaces; others are better for covering plasterboard.

Any attempt to understand how goals relate to "should" must take into account the differences between first- and second-order goals and how these two types of goals are connected to each other. It was pointed out earlier in these notes that first-order goals are diverse and specific and that the perception that they have been reached depends on sensory input – sight, sound, touch and so forth. In contrast, second-order goals are reached when a wanted emotion or feeling state is expected or hoped for, then is experienced. This occurs after a first-order goal is accomplished. It feels good when someone speaks approvingly of something we have done or made. Such approval may or may not come directly, at the time a goal is met. It may also be based on a person's memories of what parents, teachers or friends have praised or approved of in the past (here an expectation of approval is involved in the setting of a goal, and in efforts to reach it). The idea that such approval is likely to occur may be conscious or it may be thought of as (or confused with) self-approval. Some standards and rules are clearly derived from experiences not accessible to recall, like the occurrences of infancy and early childhood, experienced before event memories start to be formed, at about age three. A person may think of such standards as being connected to self-approval, but their origins are really in other-approval. A toddler feels good when it is praised or hugged after it does something which has pleased its mother, but will not later recall the specific event.

A person may experience pleasure or satisfaction if he sets himself a goal and then reaches it. These feelings do not depend on the nature of the specific goal (the winning of a race, for example), but rather on the recognition that an "I wanted that, then got that" sequence has played out (see Jean Piaget's references to the "well-known joy of being the cause" and to actions carried out, "to make interesting things last").

The preconditions for a use of "should" at a particular time or in a particular context include both inborn and acquired factors. We appear to be born with a set of intrinsic goals, which we carry, at least in latent form, our whole lives. Some of these relate more or less directly to primary sensory input; these connect almost at once to primary, first-order goals and have to do with both avoiding distressing input (pain, dizziness, nausea, a sense of smothering and overwhelmingly intense noise or light exposure and with a seeking out naturally pleasurable sensations (good tasting food, feeling warm and comfortable, sexual intercourse). Feeling state related second-order goals are inborn. They are typically experienced as a result of the reaching of first-order goals. An association chain is activated when a first-order goal is reached with the eventual result that a feeling state- holding-sending domain is stimulated and responds by sending feeling state signals (distinctive, depending on the properties of the functional domain stimulated) to consciousness.

Any individual use of "should" might be thought of as involving a sequence of elements or events. These events or occurrences rest on a background or framework of the user's inborn objectives, related to intrinsic affinities or aversions. Such sequences begin with experiences, the memories of which lead, or have led, to expectations ("when "A" happened, "B" usually happened soon afterward"), then to the development of goals (first-order goals, with their links to second-order ones) and finally, to the emergence of rules, principles or procedures which "should" be observed as particular goals are pursued.

SOURCES

The experiences of everyday life

Random House Webster's College Dictionary, Random House, New York 1991

Piaget, J., Play, Dreams and Imitation in Childhood, W.W. Norton & Co, New York 1962

Piaget, J., The Moral Judgment of the Child, The Free Press, a division of the MacMillan Publishing Co, New York, 1965

Piaget, J., The Origins of Intelligence in Children, Norton Library, W.W. Norton & Co, New York, 1963 (rev)

Dictionary of Psychology, Corsini, R, Editor, Brunner-Routledge, New York, 2002

Huber, R., Tononi, G., Circelli, C., Exploratory Behavior, Cortical BDNF Expression and Sleep Homeostasis, Sleep, 30, No. 2: 129-139, 2007

Principles of Neural Science (Fourth Edition) Kandel, Schwartz, and Jessell, Editors, McGraw-Hill Health Professionals Division. New York, 2000

The Seattle Times, Seattle

CONCERNING KNOWLEDGE & UNCERTAINTY

ABSTRACT

The development of knowledge begins with someone's receiving sensory input. What the eyes, ears and other sensory detectors detect they render into distinctive codes, which are then assembled into patterns. These patterns then go to consciousness and to memory storage (knowledge is, after all, essentially memory), after which they can be transformed and transmitted to pages as written words or from person to person in speech. Many codes are manipulated by calculation or reasoning so as to produce new knowledge (e.g. the theory of relativity). Transmission of knowledge codes requires conversion of codes from one form to another: in-brain codes to written or vocal codes, vocal ones to written words, and so forth. Errors and uncertainty in knowledge development relate to the limited capacities of human and artificial data collectors (eyes, ears, telescopes); errors in reasoning and calculation; imperfect translation and transmission of codes (brain to page, etc.); and data loss (the brain forgets, the book is burned). Despite the many opportunities for – and unavoidable occurrence of – flaws in the acquisition, storage, retrieval and communication of knowledge, knowledge has often functioned well enough to make it possible for us to have clean drinking water, plentiful power and general comfort.

INTRODUCTION

THE UNFAMILIAR FAMILIAR

The urge to exploration which led to the formulating and recording of these ruminations was aroused by several recent experiences. The first was the recognition, which came to mind for reasons I cannot now remember, that is possible to look at something familiar without attending fully to what its appearance signifies. I had come to the use of what I could see of the movement of the branches of two trees, visible from my living room window, as indicators of how strongly the wind was blowing, when deciding what coat I would wear as I prepared to leave the house. One of the trees is a conifer; the other is deciduous. One day, sometime in late spring, summer or early autumn, I realized – again for reasons I cannot remember –

that I did not know how like each other the movements of the branches of the two trees in response to the wind were in winter, after the deciduous tree had shed its leaves. I then resolved to make a point of observing the trees as the year went on, in order to settle this question to make my knowledge of how the trees were affected by the wind more accurate and complete.

The idea that there might be uncertainty about the accuracy and completeness of our understanding of familiar words or statements was brought to mind by news media reports of political candidates' pronouncements, made during the election campaigns of 2012. These utterances were, or became, familiar. They were also often contradictory, and thus could not, logically, be equally true or valid. The knowledge of at least one of any two competing candidates had to be flawed or incorrect, with respect to a subject the competitors were discussing. These experiences suggested that the meanings and implications of the words "knowledge" and "uncertainty" might be worth exploring.

KNOWLEDGE

None of the eight phrases defining knowledge in the Random House Webster's College Dictionary (see Sources) refers to the acquisition or the retention of knowledge. This omission occurs (or these subjects fail to appear) in the eleven descriptions of the word "know:" in the ten phrases explaining the word "certain;" in the six following the word "uncertain;" and the three explaining "uncertainty." These five words are all familiar; we use them often in speaking and writing without considering all their implications. Fortunately, the dictionary just referred to offers some help in indicating how an exploration of the ideas of knowledge and uncertainty might be approached, in its entry explaining another word: epistemology. This word is characterized in a single phrase, "a branch of philosophy that investigates the origin, nature, methods and limits of human knowledge."

SPECIAL TERMS & CONCEPTS USED IN THESE NOTES

INPUT

The word input refers to a signal or influence which is directed to, and arrives at, some destination. The feeling in the skin of the hand when an object is grasped is input; sound waves reaching the ear generate input. Input may be internal, within the brain. An idea or image appearing in consciousness is input with respect to a person's field of awareness (see later notes concerning the attention field). The term may also apply to artificial devices. Sound waves are input with respect to artificial sound recording systems.

FUNCTIONAL DOMAINS

The term "functional domain" is used to refer to systems in the brain having individual, distinctive properties, or in which particular functions are performed. Consciousness, the domain in which, or with which, sensations and experiences are perceived and in which actions are imagined, planned, set in motion and regulated is a functional domain (the word "action" here refers to overt, observable behavior and also covert, in-brain cognitive activity, the conscious reasoning process used to solve problems or to visualize possible results of actions being considered, for example). A functional domain may have anatomic correlates in the brain; it may also consist of a complex assembly of anatomic sites and neural networks.

THE ATTENTION FIELD (AF)

The term "attention field (AF)" is used here as an approximate synonym for "consciousness." Referring to consciousness as an attention field is intended to indicate several special features of conscious awareness. The first is that, at any particular moment, consciousness has a specific content and that the element or elements it holds is (are) limited in character and quantity (as we are reminded over and over by news reports of the consequences of distracted driving in relation to the occurrence of traffic accidents).

The second is that the content of the AF has to do with the present moment. A driver's view of the road ahead of his car and his planning and performance of the physical movements (and his sensing of those movements) used to move the steering wheel to direct the car along the road safely are elements present in the attention field (some of the driving responses may be partly out of mind and automatic, of course, with a practiced driver).

This third feature of the AF is that it is where input leads to purposeful actions, to consciously formulated expectations and to reasoning and calculation – in general to the linking of the present with expected, or hoped for, future events and experiences (an author's making up of narratives occurs in the AF).

THE PARA-ATTENTION FIELD (PAF)

This functional domain is where operations needed for programming and completing activities in progress take place in the shadows, mostly out of full consciousness. These include the retaining of information items in short-term or "working memory" while actions which depend on them are completed, like a shopper's keeping in mind the location of a cashier's check-out station while he focuses his attention on the shelves before him as he searches for a product he wants to buy (during this search he does not consciously think about where the cashier's station is). Other processes occurring in the PAF are screening and categorization of input elements so they can be recognized as in some way important (or not) or worth remembering beyond the present (or not).

Input element routing (generally following screening), like the transmitting of a bit of information to a moderately long-term or long-term memory bin ("bin" here refers to a retention site), or its retrieval from memory for presentation to the AF (including the activating of association chains needed to locate and identify items in storage so that specific relevant information can be brought to the AF) also occurs in the PAF. Other functions which must be performed in the PAF are the establishing of associations and of the chains or networks of associations just mentioned. This last set of activities is related to the development of expectations like those formed in the course of classical conditioning (the dog drools when it hears a sound it has heard repeatedly in the past just before a feeding).

CODES AND CODE UNITS

A code unit is a sign, a characteristic, patterned symbol which denotes a particular thing, idea, event or experience (including the perceptions of emotions – fear, joy, etc.). Analysis of how information that is stored in and moves through a human brain requires consideration of this idea of codes and code units. Evaluation of the functions of artificial devices which record sounds or visual images, and which can later display what has been recorded (audio recorders, video cameras and so forth) also must employ the notions of codes and code units. Written words are code units, as are the letters which make them up. Those letters, in turn are assemblies of lines of varying lengths and shapes, put together in a way which makes each letter a distinctive entity. The same idea applies to spoken words, in which, or with which, distinctive, sequential combinations of the phonemes which they contain form individual, nameable patterns.

THE ORGINAZATION OF THIS STUDY

The dictionary definition of epistemology, referred to earlier, that it is "a branch of philosophy that investigates the origin, nature, methods and limits of human knowledge," suggests a convenient way to organize these ruminations. Further, a little reflection makes it clear that this analysis is in fact an exercise in epistemology. The word "human" in the definition points to the fact that knowledge is something we humans contemplate and employ in the course of our on-going activities. A bit of knowledge might as well not exist – it really does not exist – if it does not appear at some time in the field of attention of one or more individuals.

THE NATURE, ORIGINS AND METHODS OF KNOWLEDGE

KNOWLEDGE AS MEMORY

A study of the nature of knowledge must take into account that it is a form of memory. Facts, ideas and narratives of events (and any emotions associated with them) are not bits of knowledge unless they are retained somewhere. Recognition of this aspect of knowledge leads to a need to consider some of the essential features of memory. To begin with, any memory is the result or trace, of some occurrence (or input). This trace has duration in time. The crease on a piece of paper which has been folded is a kind of memory, retained in the sheet of paper, as is a fold produced in a pair of trousers by ironing. These two examples illustrate several other properties of memory. The crease in the paper may last as long as the paper itself. On the other hand the fold in the trousers may disappear if the trousers are crumpled or shaken, or, in some cases, with the simple passage of time. Some memories are very long-lasting and some are not. Also, how long the memories last depends on how, or in what repository, they have been placed (here in paper or cloth). Any memory may be lost if what holds it is destroyed, as with burning or shredding of folded or creased paper or cloth.

Knowledge retention clearly depends on the laws which determine memory persistence in general; laws brought to our attention repeatedly in everyday life. The longer a time interval is from a memory-producing experience (input) to the time when an attempt at recall occurs, the more likely it is that the memory will have been lost, or will have been changed from something easily brought to mind at will to something only recognized with re-presentation of the input which produced it in the first place (or similar to that which first produced it). Reviewing in mind of facts just learned and rehearsal of action routines (musician rehearses a musical selection he is to play in a future concert, an athlete practices performing his role in executing a particular offensive play in football) shortens the initial input to recall time, stabilizing and extending the memory for information or action elements. The house in which I lived thirty years ago had drawers in its kitchen, in some of which cooking utensils were stored: other drawers held tableware. While I lived in that house, I could go without thinking to the drawer which contained whatever article I needed to use at any particular time (setting the table for breakfast, or getting ready to make a pot of tea). Now, I can no longer remember where in the kitchen the different drawers were located, or in which of the drawers which utensils were kept.

The intensity or the emotion-related qualities of an experience also influences its persistence in memory (consider the post-traumatic stress disorder [PTSD]). A skier who encounters an unexpected irregularity on a mountain trail he is following and falls, suffering a painful injury, is likely to remember for a long time where on the trail he came to grief. Besides the pain of the injury, the remembered emotional components of the mishap [see Kandel]) act as reinforcers of memory. The skier felt an immediate flash of fear as he fell, anger at being hurt ("that damned trail hurt me!"), and embarrassment at his having been careless and overconfident when he should have been careful and on the watch for the unexpected.

An additional aspect of relatively long-term and long term memory (ultra-short-term, short term and "working memory" have no place in this discussion – they relate to, but their elements are not part, of an individual's body of knowledge, or the body of knowledge stored in libraries and like storage sites) which calls for attention is that all memory elements with substantial persistence, when they are not on display in the attention field, are effectively unconscious. They may be in inactive storage; they may also be participants in ongoing active processes, like the screening and categorizing of input signals of the moment, or as members of activated association chains, the rattling of which brings particular memory elements to consciousness. There must be a selection process operating in the determining of which, of the many association chains or sequences held in a person's mind (associations must be themselves memories, parts of an individual's body of knowledge) is activated.

Two men are having a conversation about travel. One says, "When my wife and I were in Edinburgh last year…," and with this, his companion, who had not been thinking about Scotland up to that point, but who had been to Edinburgh, has a series of images appear in his mind's eye. He recalls the appearance of the Edinburgh castle, the castle courtyard and the view down the street running east from the castle. These images were in inactive storage at the start of the conversation until the memory bins which held them were stimulated to send them to the listener's attention field. His hearing the word "Edinburgh" sent a series of signals down a particular association chain.

Everyday experiences constantly remind us that some memories are easily brought to consciousness and that others are not. Retrieval of some items from memory (the word retrieval here does not refer to recovery of memories made less accessible by the passage of time) may be consciously, semi-unconsciously or unconsciously blocked, when hints of their possible entry into the attention field are perceived. There are some things we do not like to think about. If, in the course of a conversation, a subject distressing to one of those present is mentioned, that person may actively object to its being discussed, may try to ignore the discussion or may try to change the subject. These actions closely resemble avoidance reactions (see Dictionary of Psychology), like that of a deer grazing at the edge of a grove of trees that sees a man walking towards him through the woods and runs away. (He does not know whether or not the man may be dangerous to him, but the man is unfamiliar, and from the deer's perspective, something strange may represent a threat). The person responding to a disagreeable discussion acts as though he is making an avoidance response. What seems to be entering or approaching his attention field merits avoidance, on the basis of his past experience, even though the signals coming toward conscious awareness come in part from stored memories and the associations between them, besides being things spoken of by his companions (words, not concrete, externally presented threats).

Discussion of some aspects of knowledge as memory has been deferred to a later section of these notes. Review of some of the properties of long-term memory occurs in connection with the analysis of knowledge storage and retrieval, when memory element storage not dependent on continuing neuronal activity (depolarization-repolarization events, neuronal "firing") is taken into account. Memory retention after profound suppression of neuronal metabolism, barbiturate-induced coma, for example, and the retention of knowledge elements acquired in infancy and early childhood, when major synaptic organization is taking place fall into the class of subjects reserved for that section.

KNOWLEDGE ELEMENTS AS DISTINCTIVE ENTITIES

Knowledge is necessarily about specific things, both data, and emotions or feeling states: facts (Washington, D.C. is the capitol of the United States); action routines (how-to information: how to add and subtract, how to drive a nail with a hammer); emotions or feelings (what is to feel happy, bored or fearful)' and associations (what items belong in what class of entities or ideas, what actions lead to happiness or embarrassment). The elements of knowledge must be stored in the brain as code units. The same applies to storage sites like books or compact discs. There is no kennel in the brain from which a miniature dog is released when we hear someone say "dog." The storage of knowledge elements in the brain depends on the placement of code units in selected memory bins. Retrieval of a knowledge element for conscious consideration must make use of a stimulus-response process. An input signal comes to a memory bin, perhaps almost directly, but also at times as a result of the activation of an association chain. The target bin then sends a signal, or series of similar signals, to the attention field. When an association chain is involved (it is hard to imagine situations in which such a chain is not involved), bins lying along the chain before the target bin may or may not send code units to the attention field. Sometimes we can be aware of how a particular fact has come to mind on reflection, but this is not always true. "I can't think what made me remember how I decided to buy that shirt." In any case, bin-to-bin signal transmission occurs when an association chain is activated. Which bins are activated in any given recall-retrieval event seems to be related both to context – what activities a person is engaged in at the moment – and on the person's past experiences. Both the memory elements and their bins are individual and distinctive. If this were not the case, the right memory bin could not send the right memory element code units to the AF at the right time, in order that the evoked memory could affect the person's observable or cognitive (covert) actions.

THE ORIGINS OF KNOWLEDGE

INPUT AND INITIAL CODING

The first step in the acquisition and development of knowledge is data input. Something happens to activate a sensor – the retina, for example – and a primary code is formed. This code in combination with other sensors' codes, is eventually deposited in a memory bin. A man who has lived his whole life in a treeless desert first sees a tree (to simplify this example, we assume he has not seen pictures of trees, read descriptions of trees, or has been told about trees by a better-traveled acquaintance). His retina sends code units to the areas in his brain dedicated to receiving and routing visual input. These signals then go to the unimodal association areas nearby where patterns are formed from the code units which have come from the individual retinal sensors. These patters ultimately reach the AF (most, but perhaps not all of them) after screening in the part of the PAF devoted to classifying and routing input data. The AF and PAF together act to send the "tree" code unit to a memory bin. The man now knows that there is such a thing as a tree. The point of presenting the "man-sees-a-tree" example is to show that the acquisition of a knowledge element is a sensory event, and to suggest some of the mechanisms active during this event.

The sensors provided to us at birth are limited in number, scope and sensitivity. I can see a red object, but I cannot see infra-red radiation. We do however, come to know about things outside the detection range of our primary senses. A microscope enables us to know that organisms are composed of cells and that many cells have both nuclei and cytoplasm. On the other hand, what we learn using sensory extenders is stored on pages as visible words, pictures, diagrams and symbols; recording devices play back the audible sounds they have in storage when we press the appropriate buttons. It is clear that, as with initial knowledge acquisition (recall the "man-sees-a-tree-for-the-first-time" scenario) knowledge storage – and of course, necessarily, knowledge retrieval – depends on our genetically provided sensory equipment.

A further aspect of input related to the knowledge acquisition (and to knowledge storage and retrieval, as well) is its feeling state, its emotional dimension. The signals which come to the AF and PAF are not only those originating in the eye, the ear, the skin, and the deep tissues of the body. We experience fear, anger, excitement, satisfaction, contentment, and joy as individual, distinctive sensations, as different from each other as hearing and touch-generated perceptions. Further, we become aware of them at the same times when more directly sensory input comes to awareness. "I'm still feeling pleased that we tried out that new dessert at the restaurant last night." These feeling state messages are sent to consciousness from a group of functional domains within the brain which, when stimulated by signals which act on them as input, send code units embodying their various feeling states to consciousness. We know also, by virtue of our daily experiences, that more than one feeling state-sending domain may be transmitting signals at the same time. Excitement and joy may be blended as may be excitement and anger.

The feeling state components of units of knowledge are probably not resident only in the brains of single individuals. Knowledge held in libraries and similar storage loci may (probably does) have feeling state qualities for library users and for the libraries' staffs. What the library contains is there because the person or persons who caused it to be placed derived satisfaction in the course of discovering, developing and recording it. It remains in the library because library users find pleasure or satisfaction by reading it and contemplating it, or by using it in carrying out their activities. Knowledge elements not referred to and used (or appreciated) fade out of public awareness, like the meanings of words in the languages of groups of people which are assimilated into groups more dominant than their own (as has occurred with some Native American tribes). Some knowledge elements may be lost from libraries by disuse. "It's been years since anyone asked me to find the 1930's issues of that journal; I think they may have been thrown away the last time we reorganized the journal archives."

INPUT PROCESSING

A further source of knowledge is what our brains do with knowledge elements. Cognitive manipulations of knowledge element code units (and data processing by computers) create new knowledge. The principle of biologic evolution was discovered as a result of assessment of facts widely known before Charles Darwin began his work, although those facts had not been assembled in a way which pointed to the conclusions he formulated. The knowledge that the earth moves around the sun, and not the reverse, is another example of this kind of knowledge development.

One feature of data processing which affects the development of new knowledge is the use of measurement methods. We see a long or a short object and classify it as long or short. Our knowledge of it is augmented if we measure it with a device. Use of a ruler or tape measure makes it possible for us to say that it is four feet long, for example, and helps us to compare it with other things which may be longer or shorter (someone, sometime must have, as a result of a reasoning process, invented rulers). Whether we make rough quantitative judgments about things – "It was about as long as my arm" –

or more precise ones, as with a ruler, each assessment employs a measurement standard previously formulated and retained in a memory of some kind, in a brain, on a page in a book or in a journal.

Another aspect of input processing which shapes knowledge acquisition is the forming of associations and the retention of those associations in memory. This operation is almost certainly an early step in the development of abstract concepts, like the notions of cause and effect. "When that happens, it always hurts," or "When I see a bright flash of light in the sky, I always hear a loud noise a few moments later" (these two examples provide an illustration of one of our difficulties with forming and refining abstract ideas: the pain is likely to be a consequence of what the speaker sees, but there is no cause-effect connection between the lightning and the thunder, which are associated because both are results of a single event which produces both of them). Association connections may be built into artificial devices as well as into human brains. A computer linked to an automated telephone answering system may be programmed to respond to key words; when a caller uses one of these words, the answering system responds with a programmed message.

CONSCIOUS AND UNCONSCIOUS KNOWLEDGE ACQUISITION AND DEVELOPMENT

It is clear that we learn some things through conscious, purposeful effort and other things automatically, semi-consciously or unconsciously. The example of memory development and rehearsal which involved my knowing how to find a particular eating utensil in the kitchen drawers in the house in which I lived many years ago was based on my first having learned which drawer held which articles. This learning was largely automatic (a kind of conditioning process, perhaps; it had some purposeful aspects, of course). At a more fundamental level, much of what is learned in infancy and early childhood during what Jean Piaget calls the sensory/motor stage of development, is acquired unconsciously, in situations and by experiences the individual cannot later recall. I do not remember learning to walk, or how and when I came to recognize my parents' faces as distinctive and different from each other and from those of other people. Much of what we learn as adults, besides where to find things in our kitchens, is acquired without conscious effort or intent. We come to recognize melodies; we learn to distinguish the sounds of one person's speech from those of another.

Emotional reinforcers act to enhance initial learning, and the retention of what is learned, as was discussed earlier in the notes concerning knowledge as memory. The first hearing of a melody can produce an immediate pleasurable feeling. A person may take note of facts presented in a journal article and may be able later to remember them when they are relevant to a subject in which he interested, a subject he enjoys studying about or working with; the enjoyment here is an emotionally rewarding "feel-good" state of mind.

KNOWLEDGE METHODS

KNOWLEDGE ELEMENT DEPICTION

The initial representation of data which will be included in a person's body of knowledge occurs at primary sensors: the retinas; the acoustic nerve endings; touch, and temperature sensors in the skin, and so forth. A stimulus causes a code to form automatically and to begin its travel toward more central domains of the nervous system, including (but not limited to) the AF. The natures of the code units generated depend on the structures and inherent functional capacities of primary sensors. Including the types of stimuli to which each can respond, how the sensors create codes and the fact that, with rare exceptions, codes can travel along the neuronal pathways in only one direction. Each type of sensor's responses are dictated by what they are able to detect. The ear does not respond to light waves.

Code units move in the same way within the nervous system, whether they are sending information centrally from primary sensors or are moving signals (including both data elements and feeling states) between functional domains (from memory to the AF, for example). Moving signals must pass through neurons and from neuron to neuron. Code unit movement depends on neurons doing something, and all any neuron can do is to "fire,' to undergo a depolarization-repolarization process, during which the electrical charge difference between its outer surface (which is positive as compared with its interior when the neuron is inactive) and its interior is lost, and then restored. With each firing a wave of depolarization moves through the cell (at any one instant some parts of the neuron are depolarized and some are not) and out along its axon to the axon terminal where a neurotransmitter chemical is released, and moves to receptors on the surface of the next cell in the pathway, causing it to fire in turn (a few axon-to-target-cell transmission sites, synapses, have anatomic bridges through which signals move directly without involvement of a neurotransmitter). Signal movement is thus, as was noted earlier, one-directional (in rare cases an axon which has been artificially stimulated may carry impulses backward toward its cell body). With these facts in mind, we are brought to the conclusion that neuronal firings function as building blocks of code units. The representation, or depiction, of the elements of knowledge therefore depends on there being distinctive patterns of neuronal firing which constitute digital binary codes. Whatever may be the forms in which knowledge is stored in memory or how the different feeling states are held in reserve in their functional domains, these elements travel to the AF and PAF as such codes (or code units). Stimuli activating primary sensors or those which cause memory bins to send information to the AF, or which provoke feeling state holding domains to transmit signals representing their particular emotions must produce trains of code units. The patterns of code units formed necessarily depend on the inborn properties of the primary sensors and the emotion-related functional domains. For informational knowledge elements (rather than feeling state ones) the ultimate, basic building blocks must be codes generated in the eyes ears, skin and the somato-sensory receptors.

Time factors must affect code formation and transmission. How codes are structured and how fast they can move within the brain depend on the rates at which depolarization waves move through neurons, the time required for cell-to-cell synaptic signal transport to occur and on how many neuronal firings can occur each second (a complete depolarization-repolarization event can occur in about three milliseconds – neurons may vary in this; some neurons may be able to fire more rapidly than others). Code units must be linear. Each of a code unit's energy pulses must follow its predecessor by some typical time interval if codes are to be formed by the clustering and spacing of pulses in characteristic patterns.

Codes of some kind must be the bases for ways of depicting (and storing) knowledge outside of the brain: written words, spoken words, diagrams, symbols and pictures. However, before knowledge can be put into any of these other storage and transmission forms it must first be a digital code in a human brain. This is true even when artificial detecting and recording devices (cameras, sound recording equipment, and so forth) are taken into account. Someone's brain was the source of the ideas which led to the design and construction of such devices.

KNOWLEDGE STORAGE AND RETRIEVAL FROM STORAGE

If knowledge is to exist in any practical sense, it must be held in a storage site, from which its elements can be retrieved at need. Storage sites include human brains, libraries, computer memories, audio recording devices and albums of photographic prints. It was pointed out earlier in these notes that knowledge element storage in artificial devices can occur only after they have first been processed in a human brain. This fact calls for the consideration of two aspects of in-brain knowledge retention, related to memory functions.

First, it is clear that several different mechanisms of knowledge storage operate in the human brain. Memories of generally brief duration, like recalling where in a store parking lot a shopper's car is parked, are outside the scope of this discussion, which is concerned with memories – knowledge – having substantial duration in time. Within the class of more enduring memories are the sub-categories of memories which can be lost if they are not rehearsed or reinforced in some fashion, and those which disappear only with destruction or death of the neurons which make up the memory bins in which they are held.

Both of these two sub-classes of memory storage mechanisms ensure data retention even when the neurons (or neuronal complexes) which make up the memory bins are not firing, or are not firing at rates above their minimum baseline firing frequencies.

Since the nineteen-sixties or earlier, it has been know that deep barbiturate-induced coma, when the electroencephalogram is "flat" and demonstrates no neuronal firing, does not cause erasure of all of a person's memories. If a person who has been in such a coma (perhaps induced with the purpose of preventing brain damage by reducing brain metabolism, during treatment for life-threatening illness) recovers consciousness, he still knows who he is and can recall much of his personal history; he can recognize his friends' faces and can perform addition and subtraction (Plum, et al). At the same time, this man, like most or all of us, will have forgotten many other things, the memories of which have not been reinforced. Thirty years ago, if I had been in a barbiturate-induced coma for a time and had then recovered, it is probable that I would have been able to find a teaspoon easily in the drawers of my kitchen, although now, I have lost this capacity.

The second sub-category of long-lasting memories is made up of elements lost only if the neurons (or networks of neurons) of which their holding bins are made, die or are destroyed (Alzheimer's disease, Strokes, etc.). Much or most of such knowledge seems to be that acquired in infancy and earl childhood: how to stand and walk; the meanings and implications of "yes," "no" and "stop doing that;" what to expect when mother smiles. The period during which this kind of knowledge is acquired is the same as that when major brain maturation is taking place (synaptic pruning, myelin creation, and so forth). We are typically amnesic for events occupying this time of early life, before we reach the age of three or four (it is tempting to speculate that these basics must be learned before the event memories which make up our individual personal histories can be formed). Retention of both the potentially forgettable and the mostly unforgettable long-term memories must – on logical grounds – depend on structural changes in neurons. This has been clearly demonstrated (see Kandel's discussion of the numbers of synaptic terminals between the axon of a cell and the dendrites of the next cell in a neural pathway) in the retention or loss of memory-based responses in simple marine animals. Also on logical grounds, if a cell which is part of a center or network which functions as a memory bin dies or is destroyed, it seems inevitable that some of the data stored in that bin should be lost, even though it and its functionally associated companion neurons had been structured for retention of special knowledge elements in infancy (the idea that memory-related anatomic organization can be produced by early post-natal input is supported by the observation that in subjects deaf from birth, some cerebral cortical areas are structurally different – in the relation of quantities of gray to white mater, for example – from those same areas in subjects with normal hearing from birth onward).

An additional feature of memory element retention is the - often unconscious – determining, at the time of memory-inducing (knowledge building) experience of what data are to be retained. We take for granted that individuals vary widely in what kinds of things they know; a chess champion knows different things than a chemist. Some of these differences are due to the fact that input screening leads to different results in different people. When two people see or hear the same thing, one may remember one aspect of it and another a different one. No two people have identical past experiences, and thus there are sure to be differences between individuals in which features of a new experience are noticed (and stored in memory). A doctor, taking a walk with his wife, seeing someone walking toward him, may notice the person's limping gait but ignore what the person is wearing. His wife may or may not take note of the limp. But she is likely to be aware of the color and design of the individual's garments (this example could be presented as a consequence of the writer's occupation and of gender biases, and thus as another result of the effects of past experience – and knowledge – on current perceptions).

Quantitative factors must influence now much an individual knows (this is an issue different from the questions relative to what kinds of things he knows). A child has less stored information than an adult (or why would there be schools?). And it is at least possible that, at any age, people vary in storage capacity, in how much their memory bins can hold. Two facts support this idea. First, it is clear that infants born with microcephaly cannot learn normally. Second, intelligence tests in present use (I.Q. tests) include questions which aim to determine whether or not someone being tested has learned the things which are part of most peoples' bodies of knowledge at his age. If he has learned less than his peers, he is likely to have a low score on his I.Q. test, and I.Q. scores vary widely.

When new knowledge elements are acquired, input causes the forming of characteristic code units, the building blocks of which are codes formed in the eyes, ears, skin and other sites having sensory receptors. These code units, if they are to be retained, eventually move to memory bins, the first of which (in the history of a bit of knowledge) is in a human brain. Artificial devices have what amounts to memory bins also (video cameras, albums of photographs and sound recorders). Movement of a new knowledge element to storage outside a human brain or from an artificial memory bin back to brain occurs in several steps.

Writing information on the page of a notebook, recording knowledge elements there, involves moving data from brain to page. Multiple code unit translations and transformations occur, including the writer's visualization of how the letters and words to be placed on the page should look and selecting and performing the hand movements needed to write them.

Placement of new knowledge in the memory bin of a brain depends (must depend) on a number of automatic, partly or mostly unconscious processes. Among these are the determination of whether, or for how long, the new element is to be held in working memory (it may be needed for the performance of some activity in progress); if it is not held in working memory, whether it is to be sent to long-term memory at all, rather than being allowed to fade away; and the choice of which particular memory bin will be used. This last choice in turn relates to how a new element is to be categorized – to what already stored knowledge elements it resembles or shares features with – and how it is situated or placed with other stored knowledge elements in association chains or networks.

Knowledge retrieval events, like those of knowledge storage, require code unit movement. Memories, whether held in a brain or in an artificial device, move out of storage in response to stimuli arriving at their memory bins. These stimuli must have their first origins in the continuing stream of a person's sensory input or in some cognitive activity. Two friends meet unexpectedly in a store. The year before they both had been in a group which made a guided hiking tour through a scenic wilderness area. In the course of their conversation they were reminded of that experience, and one asked the other, "what was the name of that camp where we stayed overnight after the second day of the trip?" He friend answers, "it was called Three Corners. It was that place near the river." A chemist working on a research project needs a special reagent for carrying out his analysis. At first he cannot remember its name. Then in thinking over his problem, he remembers, first the library in which he found the journal in which he had read about the reagent, then the name of the journal in which it was described, and finally the chemical substance's name.

These examples, as presented, suggest that association chains are involved in knowledge element retrieval and that associations are involved in the choice of what memory bins are activated in any particular retrieval event. The examples do not point to all of the factors in knowledge element retrieval events, however. Those relating to emotions, to feeling states, were components in what both the hikers and the chemist experienced in their knowledge retrieval events. During the conversation between the two friends about the hiking trip, both are reminded of the enjoyment it had brought them; they enjoy talking about it. In their conversation, they re-experience some of their earlier pleasure.

FURTHER NOTES ON KNOWLEDGE ELEMENT TRAVEL AND SOME THOUGHTS CONCERNING KNOWLEDGE DISPLAY

The arrival of knowledge code unit elements at the AF-PAF complex must be followed by their persistence in those domains long enough for them to be consciously perceived and to become incorporated in some cognitive or overt activity. Retrieval of a knowledge element from memory is followed by a utilization event, even if that event consists only of its pleasurable contemplation (as in the case of the research chemist associated with the hoped-for discovery of a new scientific principle or relationship). The continuing presence of a knowledge element in a functional domain requires the operation of some special mechanism, given that travel of a code unit between domains is rapid; a code unit component is present on a portion of a neuron's cell body or a segment of its axon for only a few milliseconds. On the other hand, knowledge elements can be in consciousness for long periods. The chemist may think about his special reagent for hours at a time (during which this knowledge element may appear in and disappear from his AF repeatedly, presumably bouncing back and forth between his AF and his PAF, between consciousness and working memory). One mechanism which might produce this knowledge element persistence is suggested by the ways images on motion picture and computer screens give a viewer the impression that what he sees is continuously present. When the movement of a strip of film through an old-fashioned motion picture projector is slowed, the images on the screen flicker; further slowing of this movement leads the viewer to realize that he is seeing a succession of images, which at first had been sent to the screen in so rapidly that they seemed to be continuously present. Similar mechanisms must operate with television sets and computer displays. A like process in brain might occur in several ways.

One of these would depend on the sending by a stimulated memory bin of repetitive trains of signals directly to the AF-PAF complex, with the repetitive transmissions stopping only if a pre-programmed period of repetitive signal transmission ended (the result of the way the memory bin was structured to respond to a single stimulus) or if triggering stimuli stopped arriving at the memory bin. Another would resemble the first but would involve a stimulated memory bin sending a signal to a temporary code unit holding locus in the PAF , from which trains of signals would be sent to consciousness, while the knowledge element was being used. Several further features of knowledge element transformation affect analysis of knowledge methods. As was mentioned earlier, the history of a knowledge element includes identifying the different ways it is, or has been, depicted. Code units in brain are converted into words on a page, for example. We assume – at least, we hope – that such transformations occur without loss or distortion of the essential qualities of the knowledge elements involved. Knowledge element travel, depiction and the transformations between different forms of depiction are closely inter-related. They must take place if any given knowledge element is to be made use of by more than one individual or by any single individual at more than one time. A teacher must use words or images to tell his students what he knows. The code units in his brain must be converted into the oral and facial motor patterns needed to pronounce words or into the hand movement sequences used for writing or typing.

A special kind of knowledge element transformation is that involved in translating knowledge elements from one language to another. Such translations must be made if knowledge is to have any effects outside of the language group (nation, region, district) in which it was first acquired. A special instance of this involves the changes in how ideas and facts are expressed over time. The Canterbury Tales' events recited by a person who lived when they were written would be largely unintelligible to most present English speakers.

MORE ON MANIPULATION OF KNOWLEDGE ELEMENTS

That new knowledge is created by the use of calculation and reasoning in processing knowledge elements has already been discussed. Manipulation of knowledge also enters into several additional cognitive activities. The development of new knowledge involves the forming of cause-effect connections and the development of expectations. Associations between input elements form, often automatically and unconsciously (but sometimes consciously, as in the carrying out of research activities). The order in which bits of data come to attention can lead a person to think, "I see an 'A' so a 'B' is going to appear soon." This is especially likely to occur when the A-then-B sequence is repeated a number of times, or when 'A' then 'B' has been perceived as being especially vivid, related to some strong emotion, or bizarre feature having the quality of being contrary to ordinary expectations. The resemblance between the appearance of the "If A then B" expectation, and classical conditioning, is striking. The similarity between the forming of this kind of expectations and what statisticians do is also notable.

Other areas in which knowledge methods – knowledge in action – operate are those in which creative activity and the planning of future actions take place. Presently existing knowledge must be involved in the visualizing of things or events which have not yet existed (or taken place). An author employs his knowledge of history and of how people behave and think as he writes a novel. An architect uses what he knows about structural engineering, building materials and construction equipment when he designs a building. A politician's knowledge of practical psychology molds his campaign when he runs for elective office.

THE LIMITS OF KNOWLEDGE AND NOTES CONCERNING UNCERTAINTY

THE GENERAL CONCEPT OF LIMITS

The word "limit" may be used either as a noun (a limit) or as a verb (to limit). In either case, use of the word implies that some things or actions lie within boundaries of some sort and that others may not. The definition of epistemology quoted near the start of this discussion, indicates that a body of knowledge (or knowledge in general) cannot be infinite in extent or comprehensiveness.

LIMITS AND UNCERTAINTY RELATED TO INPUT

Knowledge acquisition is limited in a number of ways. Our eyes and ears do not respond to the full ranges of the frequencies of light or sound waves. Human and artificial data – detecting, sensory systems are selective; the eye does not respond to sound. We do not know – perhaps we cannot know – all of the phenomena which exist to which the available sensory equipment (human or artificial) cannot respond.

Apart from the selective nature of data detectors, these data – sensitive systems are potentially less than perfect, either because of the ways they are made, or because they are damaged. A retina with only two sensor units, rods or cones, can detect the arrival of light waves (even perhaps, depending on how those rods or cones are structured, a wide spectrum of light waves), but cannot detect details concerning the size, shape, or texture (or even color) of what its eye has before it. Congenital colorblindness and senile macular degeneration are illustrations of this class of perceptive limitation.

We all recognize that some of us are better at describing things in words than others and that the ability to draw pictures varies from person to person. Here are additional opportunities for the occurrence of error and uncertainty, as knowledge elements are written down, sketched or described in spoken words when individuals attempt to communicate what they know to others. Further, words and diagrams have limits with respect to what they can represent and how they portray or explain knowledge elements.

ERRORS AND LIMITS AND UNCERTAINTY ASSOCIATED WITH KNOWLEDGE ELEMENT TRAVEL

Almost any episode during which an element of knowledge travels from one site to another provides an opportunity for its distortion, modification or loss. This is the case when code units move from one functional domain to another within a brain; from one person to another; from a brain to a page, and vice versa, or from one library to another (a truck carrying a load of books is hit by another vehicle and catches fire. Many of the books are destroyed and others are damaged). Differences between individuals in life experience, education and intelligence may affect the success of the attempts of two individuals to communicate with each other. One of two persons involved in discussing a particular subject may not be paying full attention to the conversation. A lecturer or writer may choose his words poorly. There is probably no one who has not had some experience of misunderstanding and miscommunication resulting from either the use of words not well suited to express an idea or from the failure of a person trying to explain something to make it clear just what aspect of his subject he is discussing.

There are other potential mishaps in the movement of knowledge elements, both in situations in which knowledge element codes do not change during their travel (as when data move from a memory bin to the AF, or a truck carries books from one library to another) and in those which involve both information movement and code unit transformations (as when knowledge elements move from a brain to a page, or from one person to another through the use of spoken words.

UNCERTAINTY RELATED TO KNOWLEDGE ELEMENT STORAGE

Knowledge storage and retrieval from storage are additional processes, the defects and disturbances of which can lead to uncertainty about what is known. Sounds recorded on a vinyl record or compact disc may not be coded with sufficient accuracy and detail for the playing of the recorded words or music to give a listener a good reproduction of what the disc or record was intended to retain. Music played by a clarinet may have been played before the recording device, but the recording codes and systems used to convert those codes back into sounds, may (one, or the other or both together) be badly planned or organized, so that the members of an audience would be unable to recognize what instrument had produced that music.

A library may be burned; physical loss of memory codes in a brain can occur with stroke or head trauma, or piecemeal, as in Alzheimer's disease.

When knowledge codes are not physically destroyed, they can last a long time, but it is unlikely that they can be held in storage forever. Pages crumble into dust. The knowledge of word meanings and of the grammar of a language disappears when the group using that language is absorbed into some other group or when the last speaker of it dies, as with "dead languages." If the human race becomes extinct, it is likely all our accumulated knowledge will disappear if some alien species does not come upon an archive which has survived.

UNCERTAINTIES RELATED TO KNOWLEDGE ELEMENT PROCESSING

Errors and uncertainties concerning what we believe we know and what our knowledge implies can result from our having incomplete information about a subject we are studying and from how we use (and choose) reasoning or computing procedures. The sun does not go around the earth. A student just learning how to do long-division may make errors before he learns how to carry out the steps in the process correctly. Even if the student knows basic long division rules, he may make errors in performing a computation if he works hurriedly or is careless.

In complex data processing activities, errors, including the choosing of reasoning procedures to be used in solving particular problems, can be associated with the development of erroneous ideas about what propositions are true or false or what entities might exist. Some problems may arise because of the design characteristics of computing devices; some might be the results of limits of what human brains can do. Some entities or ideas for which we have names are literally inconceivable: eternity; infinity; the self; not existing.

Different conclusions are often reached in the analysis of bodies of data by different individuals, each of whom has the same information in hand and each of whom seems to have the same level of overall intelligence. The justices of the United Sates Supreme Court often disagree on points of law. Some individual differences in the results of reasoning are likely to be due to variations in the abilities of different analysts to process different kinds of information; some people have better verbal skills than others. A mathematician may be able to perform complex computations correctly but may have difficulty explaining what he does in words.

Faulty data processing occurs with several types of information manipulation. Calculating may be performed incorrectly, as in the case of the student first attempting to use long division. Analytic procedures may include logical flaws. An analytic method may not be well suited for application to a particular problem. And of course person-to-person communication of the results of a reasoning process may be imperfect. The kinds of associations formed and retained as components in in complexes or collections of knowledge elements may vary from person to person. Not everyone has the same ideas about what things belong in particular categories. Grocery store employees may place boxes of artificial sweetener on shelves in their store near sugar and baking supplies in one store; in another store they may be found beside coffee and tea.

The forming of expectations and their retention in memory is a part of knowledge element processing. This depends on the association-forming actions just noted and on the development of cause-effect perceptions arising out of repeated experiences. Both of these factors vary from person to person; both provide opportunities for the occurrence of error and uncertainty, considering how easily cause and effect relationships are confused with associations (Two events may both have the same, sometimes not apparent, cause rather than one being the cause of the other, as in the case of lightning and thunder).

A further set of influences which can produce imperfections in knowledge processing is that related to human drives and urges. Greed, a person's desire to reach a position of power or authority and the feeling of a need (which most of us seem to have to some degree) to be recognized and approved of by others may affect how knowledge elements are processed and how the results of that processing are communicated. "I believe what that man says because…." A politician may lie about what he believes to be true or he may present information in incomplete or distorted ways. An advertising message in print media or in a broadcast commercial can contain statements which – if not actually false – are still misleading. On the other hand, some information is presented in imperfect verbal or written form because the speaker or writer does not know that he is mistaken. He "just doesn't know what he's talking about."

FURTHER DISCUSSION, SUMMARY AND CONCLUSIONS

CONCERNING CODES

Knowledge is made up of individual elements, each of which is a statement about something; or a depiction of some entity. Knowledge elements are distinctive, namable entities. These bits of information must be stored somewhere; in a brain or on a page, for example. We cannot call them knowledge if they are not retained somewhere. Thus, they are essentially memories. In brain, their storage duration is determined by the factors which govern memory persistence in general. In other storage sites knowledge elements' lasting properties depend on the properties of those sites, and on how those elements are depicted and recorded. A person's acquisition and retention of bits of data involves the forming, assembly and transmission of codes. These codes are first formed when a stimulus to primary sensor or a group of sensors in the eyes, ears, nose, mouth, or skin (and the deep-in-body sensors for pain, pressure and so forth) occurs. The characteristics of the codes depend on the natures of the sensors: to what kinds of stimuli they can respond and their code-producing abilities. The properties of neurons determine that the codes representing individual input elements must be binary digital sequences. All any neuron can do is to 'fire' or to remain at rest. In order for us to perceive objects and experiences as we do, primary input codes, like those signals sent from individual retinal sensors must be assembled into patterns. An eye with just one retinal sensor does not enable a person to recognize a dog or a word on a page. Much of what we experience and remember is complex. Patterns of visual, acoustic and somato-sensory input signals are blended. Primary input has been put into patterns, first in the unimodal association areas in the brain and then in the multimodal association areas, so that what arrives at our AF is complete.

Complete knowledge elements are not just assemblies of the results of externally provoked sensory input. They also include feeling state components. The experiencing of an emotion differs from that of seeing a face or hearing a sound. The incorporation of feeling sate components into completed memories, which must be due to the stimulation of feeling state storage domains, must occur because those domains receive signals from in-brain sources. Those sources must in turn have been, or have been activated by some experiences, by some externally provoked sensory input. We do not experience every emotion or blend of emotions all the time. We feel anger or joy when some trigger leads to our awareness of the emotion associated with an event. It follows that, for us to feel and retain an emotion as part of a knowledge element, there must be the activation and operation of an association chain, the first links of which are made up of primary sensory input codes.

The codes used in the brain to depict knowledge elements must be of two general types: those involved in knowledge travel and those related to its storage. Some mention of this distinction was made earlier in these notes in references to data travel along neural pathways and to stable, long-term data retention in situations when neurons do not fire frequently (deep barbiturate-induced coma, etc.). The natures of codes involved with knowledge element travel needs no further discussion, but those code forms involved in long-term data storage deserve additional attention. A change in the neurons which make up the memory bins, or a change in their patters of interconnection must occur when a knowledge element is placed in long-term storage. The observation that – in some primitive marine animals – sensitization to particular stimuli is followed or accompanied by changes in the numbers of synaptic contacts between the axons of neurons upstream in neuronal chains and their target neurons' cell bodies or dendrites supports this inference. The ways music recording and playing devices operate indicates how this can occur in artificial devices. Changes in artificial recording devices are physical and structural, unlike the pulses of the neurons' firings, although they still may be digital. Playing a record or a disc leads to the producing of moving digital signals within the record-playing device or compact disc player and then to the generation of sound waves (also moving signals). This relationship between static and moving code units brings to our attention the fact that knowledge element storage, retrieval from storage and later travel – to the AF, for instance – requires transformation of code units from one form to another. We all assume, and hope, that these transformations occur without loss of any of the essential properties of the knowledge elements being depicted and transported. Sometimes these transformations are imperfect, and data in transit are distorted or rendered incorrect.

MORE ON KNOWLEDGE METHODS AND ERROR

Beside the signal transformations occurring in a brain when knowledge elements are being processed, as referred to in the last section of these notes, there are a number of additional transformations and translations of them in connection with knowledge movements outside, or partly outside of the brain. A person writes something he learned in a notebook or describes it in a conversation with an acquaintance.

Processes which might be thought of as among the methods of knowledge are association formation and the development of new knowledge by reasoning or calculation. Everyday life, and the disputes which arise between members of different political parties, along with the ways members of university faculties argue about subjects in which they share an interest, remind us that virtually all episodes in which elements of knowledge are transformed or are communicated provide opportunities for knowledge element distortion or abbreviation (with potential loss of important components), for there being uncertainty regarding the knowledge's correctness or completeness.

FINAL CONCLUSIONS

One of the implications of this study is that knowledge elements have histories. This principle applies both to what individuals know and to information held in books and computer memories. Several kinds of changes occur over time in the total amount of data stored and in its correctness, quality (emotional coloring, importance to the individuals or groups making use of the information, and so forth), precision and detail. Changes occur for a number of reasons. As individuals we cannot avoid having new experiences as we move from infancy to maturity and through our later lives. What has occurred with respect to the acquisition and growth of knowledge in the past argues that our exploration urge is very widespread, if not universal. It is difficult to imagine that there could be any individuals who have no curiosity at all, except during episodes of severe depression or other mental illness. We have developed new ways of acquiring and evaluating knowledge throughout recorded history. This activity has resulted in increases in the total body of knowledge and the correction of faulty beliefs, and also to the elaboration of and the discovery of the implications of, knowledge elements already in hand.

Besides the positive changes in what is known just mentioned, there are contrary tendencies which affect knowledge stores. As individuals, we forget much of what we learn. Elements of the general body of knowledge held in libraries may be (have been) lost if the libraries are destroyed. Knowledge held in an individual's brain is diminished in extent if he develops Alzheimer's disease and is lost entirely when he dies. "I'll always remember how she looked that day" is an inherently false statement. The speaker will not live forever.

Logic suggests two additional aspects of knowledge accumulation which should be considered. The first has to do with the potential extent of what might be known. That extent is itself unknown, but it is sure to be very great. We do not know how big a repository would be needed to hold all knowledge if we had the data acquisition resources needed to accumulate it. Further, we do not know whether or not the potential extent of the body of general knowledge may be infinite (whatever infinite means).

Another consideration is that knowledge acquisition takes time. There are limits to how much a person can learn in a lifetime. Each experience during which he learns something takes up a fraction of his lifespan.

Clearly there are opportunities for the occurrence of errors and incomplete formulations at every stage in the acquisition, storage, use or communication of knowledge. Some degree of uncertainty is unavoidable regarding the truth and completeness of many knowledge elements. On the other hand, this may not always be important. What we know enables us to carry out most of the activities of our daily lives successfully, most of the time. Machines do what we want them to; systems made to provide safe drinking water to the residents of cities and towns operate as they should. Perfect accuracy in measurement may not always be essential if the measurements' results fall within narrow enough tolerance limits.

One last thought: knowledge is a human phenomenon. Its extent and quality depend on the human peripheral and central nervous systems and how they respond to input. Everything we know is recorded and communicated in terms dictated by the natures of our five senses. The ultimate limits of our knowledge may well be defined by what our eyes, ears, and other sensory receptors can detect and render into codes.

REFERENCES

Diagnostic and Statistical Manual of Mental Disorders, Fourth Edition. The American Psychiatric Association, Washington, D.C., 1994

Dictionary of Psychology, Ray Corsini, Editor. Brunner-Roat-Ledge, New York, NY

Experiences of Everyday Life, widely shared

Eco, Umberto, A Theory of Semiotics, Indiana University Press, Bloomington, IN, 1976

Emmorey, K, Allen, JS, Bruss, Schenker,N, Damasio, H, A Morphometric Analysis of Auditory Brain Regions in Congenitally Deaf Adults, PNAS, Vol. 100, No 17, August 19, 2003 10049-10054

Ginsberg, H, Opper, S, Piaget's Theory of Intellectual Development, Second Edition, Prentice Hall, Inc. Inglewood Cliffs, NJ, 1979

Huttenlocher, P, DeCourten, C, Garey, LJ, Vanderloss, H., Synaptogenesis in Human Visual Cortex – Evidence for Synapse Elimination During Normal Development, Neuroscience Letters, 332, Elsevier Scientific Publishers Ireland, Ltd., 1982

Huttenlocher, P, Morphometric Study of Human Cerebral Cortex Development, Neuropsychological, 28, No 6: 517-527, 1990

Huber, R, Tononi, G, Circelli, C, Exploratory Behavior, Cortical BDNF Expression and Sleep Homeostasis, Sleep, 30, No 2: 129-139, 2007

Kandel, E, In Search of Memory, WW Norton and Co, New York and London, 2006

Kandel, ER, Schwartz, JH, Jessel, TM (Editors), Principles of Neural Science, 4th Edition, McGraw-Hill, Health Professionals Division, New York, 2000

Lewis, CI, Langford, CA, <u>Symbolic Logic</u>, 2nd Edition, Dover Publications, Inc., New York, 1932 and 1959

Plum, F, Posner, JB, <u>The Diagnosis of Stupor and Coma</u>, 3rd Edition, FA Davis Co., Philadelphia, 1982

<u>Random House Webster's College Dictionary</u>, Random House, New York, 1991

Shewhart, WA, <u>Statistical Method From the Viewpoint of Quality Control</u>, Dover Publications, Inc., New York, 1939

Silver, S, Miller, WR, <u>American Indian Languages</u>, University of Arizona Press, Tucson, 1997

Volps, JJ, <u>Normal and Abnormal Brain Development</u>, Clinics in Perinatology, Vol 4, No 1. March 1977

AFTERWORD

This collection came about because, having been convinced over time and by study, of their worth, it didn't seem right for others who love to think, study, and question to have to wait for them to published piece by piece over a long period of time. The author feels that every one of them have places that could be improved; maybe that is true, but they stand up by themselves quite well as they are. As it is, the essays clamored to be written and now are clamoring to be published and heard in a wider world. Though publishing the essays did not start out to be a goal, it has come to be part and parcel of the whole process from observation to conclusion; after all feedback is necessary.

The urge toward perfection and unflawed knowledge is a strong one; it is most likely one of the foremost reasons these essays exist. For along with greater knowledge and perfect understanding comes a much better understanding of oneself and others; which knowledge leads to comfort and contentment. Choosing to face a challenging task is the most important thing we can do – without a challenge, what is there to motivate us? Although the sheer fun of playing with words in any number of ways adds to the appeal. And having lively discussions that can lead anywhere. Mostly these essays came about, however, because their subjects are just, plain interesting.